The Chemostat

Chemostat and Bioprocesses Set

coordinated by
Claude Lobry

Volume 1

The Chemostat

*Mathematical Theory of
Microorganism Cultures*

Jérôme Harmand
Claude Lobry
Alain Rapaport
Tewfik Sari

WILEY

First published 2017 in Great Britain and the United States by ISTE Ltd and John Wiley & Sons, Inc.

ISTE Ltd
27-37 St George's Road
London SW19 4EU
UK

www.iste.co.uk

John Wiley & Sons, Inc.
111 River Street
Hoboken, NJ 07030
USA

www.wiley.com

Library of Congress Control Number: 2017938650

British Library Cataloguing-in-Publication Data
A CIP record for this book is available from the British Library
ISBN 978-1-78630-043-0

Contents

Introduction

The chemostat is an experimental device invented in the 1950s, almost simultaneously, by Jacques Monod [MON 50] on the one hand, and by Aaron Novick and Leo Szilard on the other hand [NOV 50]. In his seminal article, Monod presented both chemostat equations and an example of an experimental device that operates continuously with the aim of controlling microbial growth by interacting with the inflow rate. Novick and Szilard, for their part, proposed a simpler experimental device, one of the technical difficulties at the time being to design a system capable of delivering a constant supply to a small volume reactor. Originally named "bactogène" by Monod, Novick and Szilard are the ones who propose the name chemostat for *chemical [environment] is static*. It is used to study microorganisms and especially their growth characteristics on a so-called "limiting" substrate. The other resources essential to their development and reproduction are assumed to be present in excess inside the reactor. It comprises an enclosure containing the reaction volume, an inlet that enables resources to be fed into the system and an outlet through which all components are withdrawn. This device presents two main characteristics: its content is assumed to be perfectly homogeneous and its volume is kept constant by the use of appropriate technical devices making it possible to maintain continuous and identical in and outflow rates. Its reputation is mainly due to the fact that it is capable of fixing the growth rate of the microorganisms that it contains at equilibrium by means of manipulating the inflow supply. First used by microbiologists to study the growth of a given species of microorganisms (referred to as "pure culture"), its usage greatly diversified over time. In the 1960s, it became a standard tool for microbiologists to study relationships between growth and environment parameters. In the 1970–1980s, it would become the focus of a strong interest in mathematical ecology even though it was somewhat neglected by microbiologists. This was mainly because at the time, the attention of the latter was attracted by the development of molecular biology approaches for the monitoring and understanding of microbial ecosystems. Studies on the competition of microorganisms rekindled interest among researchers for the chemostat in the 1980s, especially in the field of microbial ecology. It is not until the 2000s and the advent of the postgenomic era, which requires knowledge and fine control of reaction media, that a renewed interest was really observed for this device among microbiologists. It is used nowadays in scientific areas related to the acquisition of knowledge that is both fundamental, such as ecology or evolutionary biology, and applied such as water treatment, biomass energy recovery and biotechnologies in a broader sense.

The chemostat has not only been the subject of numerous publications, but also of several books essential in the field of mathematics. A question can be legitimately raised about what additional work can be done concerning a device that ultimately is very simple in principle. To this question, we can provide the several following answers.

The main source of uncertainty when a biological process is being modeled lies in modeling the growth rate of microorganisms. Bearing in mind that practitioners' concerns must be addressed, it is based on this fundamental question and from an applied point of view that we have built this book. We do realize that the analytic expression of a growth rate is only an approximation thereof and that the properties of the model should not depend on this expression; that is the reason why we will introduce general models involving different growth models which we will subsequently specify. In particular, we study the influence of the type of growth function being considered on the outcome of a competition between several species. Adopting an increasingly complex approach, and after a general introduction which is covered in the first chapter, the second chapter naturally focuses on the growth of a single species of microorganisms on a resource. The properties of this model are analyzed for the three most important classes of growth rate encountered in biotechnology, namely, those limited and/or inhibited by substrate and those known as density-dependent, in which the growth rate does not depend on the resource only, but also on the density of the existing microorganisms. This first situation becomes more complex in Chapter 3 where we address the case of several species competing on a single resource, when growth rates are resource-dependent only. In particular, the competitive exclusion principle at equilibrium is therein exposed, of which demonstrations are given relating to the existence and the local as well as global stability of the equilibria of the system. The fourth chapter specifically addresses the case of several species competing and coexisting on a resource when growth rates are density-dependent. With the study of this type of model being significantly more complex from the mathematical perspective, it emerges that resorting to numerical simulation tools is a means to bring forward the diversity of the situations encountered. Finally, the final chapter addresses models enabling the consideration of the spatial structuring of microorganisms into several classes, here both in planktonic form and flocs or in biofilms.

We restrict our observations to situations in which a single limiting resource is considered. In a real situation, it is obvious that this will not be the case. Nonetheless, the reasoning needed for the study of these more complex situations will always be carried out based on the tools that we are introducing in this book. This publication is written to allow a linear reading in the order of the chapters that constitute it. However, according to the degree of detail, we propose several times that the reader overlooks certain passages, bearing in mind the possibility of returning to them later on, without confusing the progression during reading.

This book is above all dedicated to engineering students and PhD students wishing to study the techniques for the analysis of dynamic systems related to biological systems used in biotechnology and, in particular, to the chemostat (homogeneous system continuously operating). The primary concern is to address the challenge of studying the qualitative properties of a model already available and dedicated to formalizing a situation of interest. The question of confronting this model to data falls outside the scope of this book. We hope that the educational efforts achieved can make its reading accessible to the greatest number of people, including biotechnology students and not only mathematicians. In particular, important techniques are specifically detailed, whereas the elements requiring more significant developments or secondary significance are proposed as exercises. Furthermore, their solutions are included at the end of the book.

A rather significant appendix (Appendix 1) is dedicated to the theory of differential equations. Strictly speaking, this is not a course but a refresher of the principal notions and results to which we will refer. The reader equipped with the knowledge of a preparatory scientific class or a Bachelor of Science should be able to follow. The book contains a very large number of figures that most often will benefit the student when viewed enlarged (when reading the electronic version).

1

Bioreactors

1.1. Introduction

1.1.1. *What is a bioreactor?*

A bioreactor is an enclosure containing a nutrient medium consisting of a cocktail of various molecules − referred to as "substrates" − upon which one or more populations of microorganisms grow, and as such the set of these microorganisms is called "biomass". Bioreactors are used to perform operations for transforming matter through biological pathways, most often accompanied, but not systematically, by the increase of biomass in the reaction medium. Microbiology teaches us that only soluble substrates − that have created chemical bonds with water molecules − are available for the growth of living cells. Within the context of this book, from a formal point of view, a biological reaction will therefore describe the transformation of elements existing in the medium in soluble form into a solid form, biomass and possibly into a certain number of metabolites and/or gas. However, it may happen that a number of resources are present in solid form. A so-called "hydrolysis" step is necessary to transform this solid substrate into a soluble form assimilable by microorganisms. The conception that claims that a microorganism "grabs" molecules passing nearby is somehow a figment of the imagination. For example, it is accepted in soils that most microorganisms, whether they be mobile or not, excrete enzymes around them and recover the nutrients that reach them by diffusion. It is therefore essential to properly distinguish between the different processes involved before initiating the modeling of phenomena as complex as the degradation of a set of substrates by microorganisms. This is the subject of this chapter that describes the most important processes involved in this type of matter transformation and the systematic approach widely adopted in process engineering to model them. In what follows, we will only cover microbial ecosystems that are implemented in most bioprocesses.

1.1.2. *Classification of biological reactors*

In process engineering, bioreactors are first classified according to their mode of operation, in other words the way in which they are supplied with matter, and depending on whether microorganisms are free in the medium (so-called "planktonic" organisms) or fixed on a support; the latter could itself be fixed or mobile.

As a result, it is possible to distinguish continuously-fed systems, systems whose supply is semi-continuous and those operating in closed mode. In continuous reactors, the reaction volume remains constant, in- and outflow rates being identical. It is the most commonly used operating mode in industries aiming to process a large amount of material arriving continuously, as it is the case, for example, in the treatment of water by biological means. We will often refer to this mode in this chapter insofar as this is one of the most significant industries in terms of quantities of processed materials. Semi-continuous (or *fedbatch*) reactors are systems whose inflow rate is not zero but whose outflow rate is zero. In such a system, the reaction volume is thus increasing over time from a minimal to a maximal value. This type of system is particularly suitable for the production of biomass as the amount of substrate can be supplied according to the specific needs of microorganisms. It is also used when the risk of inhibition due to the substrate accumulation or a metabolic intermediary in the medium is present. Depending on the physiological state of microorganisms, it is then possible to decrease, or on the contrary, to increase the amount of resource fed into the reactor. Finally, batch mode — or reactor — designates a closed system in the sense where there is neither supply nor withdrawal of the system: substrates (the different nutrients necessary for the growth of microorganisms) as well as the inoculum (biomass) are introduced at the initial time. Therefore, the reaction volume of the system is constant over time (if possible liquid-gas exchanges are neglected) and the reaction takes place up to the moment when it is measured (or considered) that it has completed. This operating mode is widely used in agri-food, pharmaceutical and chemical industries, notably for the production of molecules with high-added value, and more generally in cultures in which the risk of contamination through the feed is high.

Since biomass is the catalyst for reaction, the effectiveness of a biological system will be all the more significant when the substrate necessary for its growth is in an appropriate form (this is referred to as biodegradability) and accessible (so-called accessibility). The homogeneity of the medium as well as biomass and resource densities will consequently play essential roles in the operation of these systems. In order to maximize the concentration of the existing biomass — but mainly to facilitate the separation of the biomass from the residual reaction medium (in other words, to facilitate the separation of the liquid and solid phases of the medium) — it is possible to resort to using a support upon which biomass will tend to settle in the form of "biofilms"[1]. In laboratories, even today, many engineers are testing the effectiveness of all kinds of fixed or mobile supports and are studying the properties of associated processes. In fact, it is essentially based on these considerations — relating to the feeding modes of reactors and to the manner in which biomass is retained within the system — that different technologies of reactors have been proposed. Finally, the last major element for the classification of bioprocesses is linked to the same biological processes that condition bacterial growth. It designates the set of conditions that must prevail in the medium to enable the growth of microorganisms (this is often referred to as "environmental conditions"). They are essentially ecosystem-dependent. In the next section, we review a certain number of concepts that are necessary to understand the formalization of the model of the chemostat, which we will next present in several ways in the book.

1.1.3. *A brief reminder of microbiology*

To grow and multiply, a (micro-) organism needs a multitude of elements, including some at trace level. A natural manner to classify organisms is to refer to the mechanisms that they implement to capture the matter necessary for their growth and to produce their energy. When we concentrate on the major factors influencing growth, a source carbon and an energy source are in effect essentially needed. Carbon is found in two basic forms in nature: organic or inorganic. For simplicity, assume here that organic carbon is the carbon produced by living entities, inorganic carbon designating the CO_2 in its different chemical forms (carbonic acid, bicarbonate, dissolved CO_2; we will later on return to these elements to talk about the mutual influence of biological reactions and chemical balance of the medium, which are fundamental factors affecting the growth of microorganisms).

There are essentially two sources of energy: light and chemical energy. Organisms that extract their energy from light are called phototrophs, those who take it from chemistry being called chemotrophs. Regarding sources of carbon, organisms being able to utilize organic matter are called heterotrophs, those using CO_2 are known as autotrophic. By combining the carbon and the energy source being utilized, four major classes of microorganisms can then be defined:

– chemoautotrophs utilize CO_2 as a carbon source and derive their energy from the consumption of inorganic substrates;

– chemoheterotrophs utilize organic matter as a source of energy and carbon;

– photoautotrophs utilize CO_2 as a carbon source and derive their energy from light;

[1] Barring a few exceptions, it may be helpful to recall here that the natural form of microbes in the medium involves that they precisely structure themselves into biofilms and flocs, the planktonic state being specific to certain species.

– photoheterotrophs utilize organic material as source of carbon and light as energy source;

In environmental biotechnology, chemoheterotroph bacteria have been particularly studied because they degrade organic matter and are not constrained by light, which is difficult to evenly diffuse inside reactors of large volumes comprising high cell densities. The metabolism of these organisms enables energy to be produced and the final products of the degradation are CO_2 and water. The other bacteria widely studied in the environmental field are chemoautotrophs because they are involved in the nitrogen cycle. They derive their energy from the presence in the medium of molecules such as ammoniacal nitrogen or nitrite and their source of carbon is CO_2.

Why have we presented these various notions in a book dedicated to the chemostat? Based on the previous few examples, we will show in the next section that the apparent complexity of the functioning of microbial ecosystems can be formalized in a simple and natural manner at the populations level, regardless of the type of metabolism implemented by the microorganisms.

1.2. Modeling of biological reactions

1.2.1. *Regarding the state variables of the model*

As indicated at the beginning of this chapter, a chemostat consists of an enclosure in which resources are provided by means of a feed and from which the reaction medium is withdrawn at an outlet (Figure 1.1). Assuming that all the elements necessary to the growth of the microorganism that will develop inside the reactor are present in excess at all times, the velocity at which it could develop would be limited only by its own ability to "ingest" substrates and at the velocity with which the medium is renewed. Assuming that its velocity of growth at time t depends on its own density (which means that with a biomass concentration twice as high, the velocity would be twice as significant), its growth would be exponential until the resource or space start lacking. In practice, this situation can only be observed during the transitional phase and, if it is not space, a resource at least, said to be "essential", always becomes scarce and limit growth (hence its name of "limiting" resource or substrate). In microbiology, this is most often the carbon or nitrogen source but, assuming they impact on the growth velocity in a similar manner to a substrate, it can equally concern light or the concentration of oxygen, that is to say, the source of energy. To achieve modeling of the chemostat, we will assume that all the necessary conditions for the growth of a microorganism are gathered except for one. In other words, with the exception of an essential substrate, it is assumed that all the nutrients needed for growth are present in excess.

It should be observed that the terms s and x both commonly denote substrate and biomass as well as their respective concentrations (thus in mass per unit volume). In this respect, let us clarify for a moment what "biomass" designates here. A microorganism, such as any living organism, is mainly composed of water. Therefore, how can a concentration of microbes within a biological reactor be characterized? Besides the available measures to which we will return to very quickly, we can address a quite fundamental question – and still subject to debate among scholars – namely the most appropriate terminology to define a set of microbial individuals that perform the same function within an ecosystem. Historically, the term "pure culture" is used to designate microorganism growth within a reactor that has been inoculated with a colony originating from a unique microbial cell. The reactor is then designed such that the ecosystem cannot be contaminated by another microorganism. In this case, this is known as "microbial strain". After numerous subcultures, some evolution will, however, be noticed in the genome of daughter cells. From the moment these differences become identifiable, but remain minor, it is then referred to as "species", a term borrowed from the classification of macroorganisms. Despite being correctly defined in the context of sexual reproduction, it has long been used in the microbial world even if modern methods of molecular biology, which have particularly highlighted that "two clones were never completely so", appearing to raise once again the issue of the existence and relevance of the notion of species in microbes (see [WAD 16]).

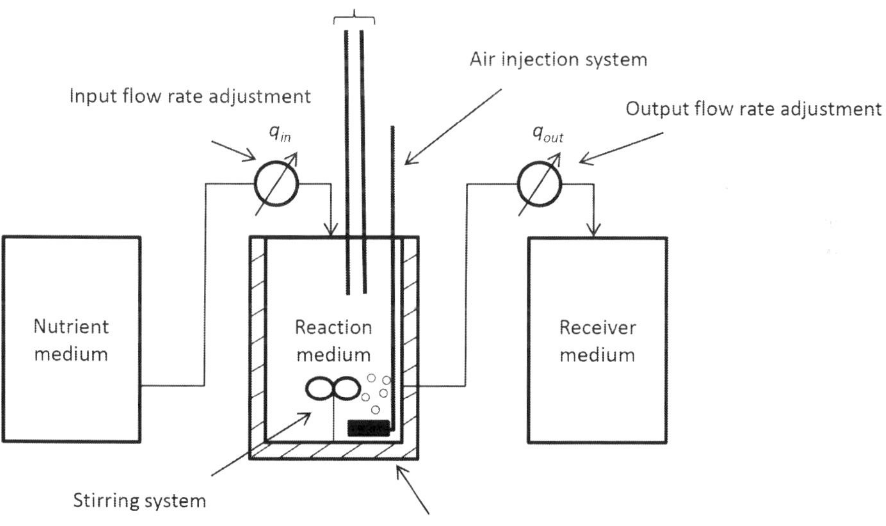

Figure 1.1. *Schematic representation of a chemostat*

The somewhat opposite concept to "pure culture" is that of "mixed culture" where existing microbes can be of different natures and perform very different functions within the ecosystem. In reality, this is of course the case of any natural ecosystem in which a large number of microorganisms live with very different characteristics. It is interesting to specify here for neophytes that a larger number of differences might be identified between the genomes of two microorganisms than between a human and a goldfish. The purpose of this book is not to address these expert issues, and we will merely define the terms, used interchangeably in this book, strain, species or even biomass (although this last term will preferably be used to designate the set of the active population): namely a set of microbes performing an identical function with respect to one or more limiting resources in a growth medium. Nonetheless, it is important to be aware that this relates here to a crucial point of debates centered on the chemostat and its modeling in particular. As a matter of fact, we will further discuss in this book questions related to diversity and coexistence using as a definition of species the one we just have given. Therefore, as we have just clarified, the function of a species is established once and for all in our models.

In fact, numerous ecologists remain convinced that modeling works, when they are applied to issues notably concerning diversity and do not raise much interest as long as the notion of evolution (the way in which the functional capabilities of the species change) is not taken into account. Without getting into very detailed arguments, we will first address these reservations that the chemostat − and its model − have initially been proposed by biologists based on technical developments and experimental difficulties: Monod and Szilard were indeed biologists. In fact, the chemostat is not − not only in any case − a figment of the imagination of mathematicians. Then, despite the fact that the models that we use, and which are to be seen here as models "to think", are necessarily very simplified visions, or even naive, of reality, it does not lessen their ability to provide a means to formalize knowledge. Their study allows the testing of assumptions, especially concerning the mechanisms involved in the reaction medium, and as a result they constitute a terrific communication tool, for exchanges and reflections between researchers whether within but even more outside of their disciplinary fields. It is too often forgotten that a model is primarily a tool for the formalization of knowledge and − not only − a tool for prediction as a means to "stick to data".

This issue of the characterization of biomass immediately recalls the technologies available for characterizing the populations of microorganisms. From the outset, these techniques raise a number of issues that sometimes appear very complicated to answer. For example, if a quantification technique requires a significant volume of samples (compared to the volume of the reactor), it could be asked to what extent this sampling will disrupt the operation of the system. What can be done to quantify the existing biomass? Let us immediately clarify that we exclude from our remarks the case of the characterization of biomass structured in flocs and in biofilm that raise technical questions which largely fall beyond the issues addressed in this book. When they are free in the medium (we recall that this is referred to as "planktonic" organisms), we know how to count cells, to characterize a structuring according to the population size or even to weigh matter by means of various techniques. It is also possible to resort to molecular biology in order to quantify the abundances of genes. However, since microorganisms reproduce through cell division, they are not all in the same physiological state at a given time. In addition, it is particularly complicated, not to say impossible, except in very particular study situations, to know which ones are active — in the sense where they participate in the observed consumption of substrate — and which are not. It should be noted that the very notion of dead microorganism must be specified since a cell may very well be undivided but inactive and that a lysed cell can release active enzymes in the medium.

These remarks have led scientists to quickly ask themselves the question about knowing what was the most appropriate characterization for measuring biomass. In the majority of cases, it implies that the effects of mortality have thus to be ignored based on the fact that, considering very large populations, their consequences are negligible on the manner in which biomass as a whole behaves with regard to a limiting resource within a continuously operating reactor. Another way to state it is that in certain situations, biomass mortality will be neglected compared to the momentum generated by the renewal of the medium.

Getting back to the measures available to quantify a concentration of microorganisms, we may make use of the concentration of "volatile matter". Considering now the technical details, the concentration of volatile matter in the medium is obtained in the following way: first, the liquid and solid phases of the sample are separated (for instance through centrifugation). We then proceed to a first weighing of the pellet (solid matter): in this way, the "concentration of solids or solids in suspension" is obtained. Assuming that there is no organic matter in solid form in the medium or in the feed, it should be noted that this concentration may be a first characterization of x. Nevertheless, this estimate can be refined by excluding from this quantity the set of minerals, assimilated to resources rather than to biomass: to this end, the sample is oxidized (through combustion) such that all of the organic matter is oxidized into gas composed of CO_2 and H_2O and the remaining matter is weighed. The difference between this last measure and solid matter gives us the volatile matters often assimilated, in biotechnology, to biomass. The fact that only part of the matter is active and the rest is the reflection of dead cells is usually overlooked due to the remarks formulated earlier. An alternative very often used if the medium is neither colored nor too loaded with suspended matters (with solids) is to measure the turbidity of the medium (its "blurred" character). This is an optical measure widely utilized in biotechnology because it is very simple to implement, even continuously with "turbidimeters" directly immersed in the medium. Nonetheless, any wastewater treatment applications, excessively loaded with matter in suspension, and more generally all those in which the environment influences the measure independently of the biomass concentration (a substrate itself can disturb the medium) cannot use it. It is also possible to adopt a formalism in which biomass and substrates are quantified in the same unit. This is how water treatment engineers proceed for whom biomass and substrates are characterized based on using the chemical oxygen demand or COD. The COD is a standard measure of pollution as well as a means for easily establishing overall matter balances. Furthermore, the incoming COD is subdivided into different compartments, all quantified *via* the same measure. Finally, it is possible to use a number of means for indirect measurement (conductivity, infrared, photometry, etc.), keeping in mind, however, that all these devices can be very useful for the systematic online monitoring of biomass. However, they have to be calibrated and therefore it is necessary to resort to a standardized concentration unit, which brings us back to the original question. First, it can be retained therefrom that adopting a unit for biomass is a challenge in itself and depends on the means available, not only human resources and equipment but also financial.

Fortunately, we are equipped with some background on the usage of these different techniques whose advantages and limitations are now quite clear. The purpose of our book is not to conform models to reality but to study their properties by means of mathematical analysis. Moreover, the key point to remember is that, later in this book, appropriate units will always be used such that up to a performance coefficient, what we claim remains valid for what has been defined as being s and/or x.

1.2.2. *Biological processes and reaction scheme*

The reaction scheme describes the desired processes to be formalized during a given biological reaction. It is inspired by the way in which reactions in chemistry are represented. A biological reaction can be seen as the transformation of matter catalyzed by the presence of a microorganism. A very intuitive manner to represent the growth of a microorganism x on a substrate s is to use the formalism of chemistry for an irreversible reaction, namely:

$$s \rightarrow x \tag{1.1}$$

that schematizes a transfer of matter from s toward/into x.

Here, the notations x and s have no unit. They are used to describe a qualitative transfer of matter, in some way a sense of reaction. In particular, there is no notion of balancing the reaction: we merely define a scheme where a matter compartment in a given form (here s) is converted into another compartment (x). This is the reaction scheme. This information can be completed by the fact that, for example, this reaction is accompanied by the production of CO_2. In this case, a term will be added to the right:

$$s \quad \rightarrow \quad x + CO_2 \qquad\qquad [1.2]$$

To be a little more specific, we can still get closer to the formalism of chemistry by indicating reaction yields and kinetics. However, we should bear in mind, once again, that yields are not originating from balancing the reaction in the sense of chemistry, but from the knowledge made available through *ad hoc* trials by end users: with regard to biomass growth, we would have previously established what is meant by "biomass" and that, according to this definition, "such amount of substrate is necessary to produce such amount of biomass". We then obtain the following scheme:

$$k_1 s \quad \overset{\rho(.)}{\rightarrow} \quad x + k_2 CO_2 \qquad\qquad [1.3]$$

which reads "the growth of a biomass unit is achieved from k_1 substrate units and is accompanied by the production of k_2 units of CO_2 at velocity (or *rate*) $\rho(.)$" (in mass per volume and per unit of time).

Let us for a moment consider the notion of yield. Assume that an organism grows in aerobiosis, on glucose (with chemical formula $C_6H_{12}O_6$) — it is thus a chemoheterotroph microorganism — and that glucose is in limited quantity. If we utilize elemental composition data from microorganisms "in the main elements that constitute them on average" (for example here in carbon, hydrogen and oxygen), it will be then possible to assimilate the biomass to a chemical formula and to use this time the formalisms of chemistry to calculate the values of these yield coefficients. Therefore, model average compositions of microorganisms can be found in the literature (see [TRU 09]). The reader should also notice that, in the formalism of the previous example, the yields can be normalized by defining k_1 or k_2 as equal to the unit, such that only a single yield parameter is considered. This parameter will be denoted by $\frac{1}{Y} = \frac{k_1}{k_2}$ (or its converse depending on whether we are considering the "consumed substrate or CO_2 produced"), which will limit the number of parameters of the model. Moreover, this is the notation that we will use in section 1.3.1. At this stage, there is no need to clarify what exactly $\rho(.)$ is and we will merely keep in mind that it is homogeneous to a velocity.

In the most elementary form previously used, the reaction scheme is capable of specifying that only a single biological process — here growth — is involved. However, according to the formalization being considered, the processes to be modeled can be multiple: biomass can, for example, be subject to "mortality". In this case, the previous scheme has to be completed by a second equation:

$$k_d x \quad \overset{x}{\rightarrow} \quad x_d \qquad\qquad [1.4]$$

which describes the fact that the "active biomass" compartment feeds a "dead biomass" compartment (denoted here by x_d) with a rate k_d.

In addition to mortality, the processes of interest in microbiology are numerous. We give hereafter some examples of processes which have been considered in the literature dedicated to modeling in microbiology:

– maintenance is the use of part of the resources for the supply of energy to biomass. It is usually modeled by adding a negative term (also known as "well") in the equation of the resource without modification of the dynamic equation of biomass, which correctly describes a disappearance of substrate without microbial growth;

– although different in terms of process, the definition of maintenance is sometimes confused with a mechanism called "endogenous respiration" that describes a situation in which the biomass itself becomes its own substrate. In other words, it concerns a process for recycling the matter based on biomass mortality. To model it, a negative mortality term can be added to the equation of biomass associated with a positive term in the dynamics of the substrate. Since there is necessarily less consumable substrate that appears than biomass that disappears in a given unit, this term is sometimes not added to the biomass due to the fact that, by extension, mortality and maintenance are then modeled by the same modifications to the equations. This situation is an excellent example of the contribution of modeling to the formalization of microbial dynamics: if general definitions are considered excessively and not formalized in the form of equations, it may result that some processes of different nature, however, appear as identical. From the moment an equation is stated, it is important to properly define the assumptions that allow us to write it. If the maintenance describes a substrate consumption without bacterial growth, then only the introduction of a negative term in the substrate is consistent. If maintenance describes a way for biomass to maintain itself in the absence of substrate, then it concerns the recycling process of nutrients modeled by the simultaneous addition of a mortality term in the dynamics of biomass and a source term in the dynamics of the substrate;

– flocculation is the structuring of biomass into flocs. As noted above, microorganisms naturally tend to agglomerate in biofilms and flocs. A more realistic description than the one that only considers planktonic microorganisms, thus involves adding a compartment of structured microorganisms to the model, for example in flocs;

– as mentioned in the introduction of this chapter, it may happen that the limiting resource in the feed may not be accessible to microorganisms, for example because it is found in solid form. In this case, it is necessary to add a hydrolysis step describing the manner how − velocity and yield depending on the conditions of the medium − this matter compartment feeds a biodegradable and accessible substrate compartment.

Keeping these general points in mind, let us now address the modeling of a simple biological reaction.

1.2.3. *Chemostat equations*

Prior to establishing the chemostat equations as such, we specify a few conventions for the notations that we will use throughout the book. To do this, it is helpful to recall here a common notion of automatic control, namely that of a system input. If we define the system of the chemostat as comprising an enclosure as well as two supply and withdrawal devices notably, including pumps, that allow the user to set inflow and outflow rates, it then becomes possible to identify variables allowing us to interact with the system. These variables are precisely the two feed and effluent flow rates and eventually the concentration of feed, of which we can picture that an *ad hoc* device allows the user to fix it to a given value. The dynamics of the volume of the reactor and the biomass and substrate concentrations inside the chemostat (constituting the state of the system) are then governed by non-independent dynamic equations insofar as variations of the input variables will lead to changes in the system state variables. By convention, for these input variables (here, the two feed and outflow rates in addition to the concentration of substrate in the feed), we will use terms in uppercase for constant values and in lowercase when these variables are likely to vary over time.

To establish the equations of the chemostat, we use here the usual formality of process engineering by directly applying a mass balance to s and x. Consider an enclosure of volume v equipped with an inlet — the supply of the reactor referred to above — and an outlet through which the reaction mixture can be withdrawn. In order to be as formal as possible, we assume that this facility is equipped with all the necessary control devices so that the mixture is homogeneous. In addition, it is also assumed that from a reactive point of view, environmental conditions (in particular temperature and pH) are constant (in the sense that they are not the reason for the variations observed in concentrations of interest).

From now on, s_{in} (or S_{in}) should denote the concentration of s in the feed, q_{in} and q_{out} (or Q_{in} and Q_{out}) the inflow and outflow rates (in volume per unit of time) and $Y(\cdot)$ the yield of the conversion of substrate into biomass (in mass of substrate consumed per mass of biomass produced). In the great majority of articles available in the literature, yields are constants. However, to explain oscillations that appear in some experiments, some authors have proposed regulatory mechanisms resulting in considering yields as functions of substrate and/or biomass concentration. In most cases, we will therefore denote yields by $Y(\cdot)$ in this chapter. The reader should also notice that we have used coefficients k_i in reaction schemes to not show terms in $\frac{1}{Y}$. According to each one's modeling cultures, conversion yields are thus differently denoted and are thus homogeneous either to a mass of substrate consumed per mass of biomass produced, or conversely, that is to say to a mass of biomass produced per mass of consumed substrate. The latter are by definition non-zero, the notation does not matter in the present case and in the following we will opt for the notation in use in the field of wastewater treatment engineering, namely a term in $\frac{1}{Y}$ in the dynamics of s. Now, consider the simpler reaction scheme [1.3] with $k_1 = \frac{1}{Y}$ and focusing on the dynamics of s and x. Let us achieve a mass balance according to which, for a period of time dt, the variation in the mass of an element (here s and x) is the result between the mass of that element that has been brought into the system added to the produced mass of this element minus the consumed quantity minus the extracted quantity. If this principle is applied to the biomass and to the substrate in the reaction volume v of the reactor, the following equations are obtained:

$$\begin{cases} \dfrac{dv}{dt} &= q_{in} - q_{out} \\[2ex] \dfrac{d(sv)}{dt} &= q_{in}s_{in} - q_{out}s - \frac{\rho(.)}{Y(\cdot)}v \\[2ex] \dfrac{d(xv)}{dt} &= \rho(.)v - q_{out}x \end{cases} \qquad [1.5]$$

Without loss of generality, we will of course be cautious to only consider rates q_{in} and q_{out} that maintain v positive. Noting that $\frac{d(uv)}{dt} = u\frac{dv}{dt} + v\frac{du}{dt}$, we can express the dynamics of the concentration of x and s that are more easily manipulated in chemistry than masses inasmuch as it is desirable to reason regardless of the volume of the reactors. It thus yields:

$$\begin{cases} \dfrac{dv}{dt} &= q_{in} - q_{out} \\[2ex] v\dfrac{ds}{dt} &= q_{in}s_{in} - q_{in}s - \frac{\rho(.)}{Y(\cdot)}v \\[2ex] v\dfrac{dx}{dt} &= \rho(.)v - q_{in}x \end{cases} \qquad [1.6]$$

or still:

$$\begin{cases} \dfrac{dv}{dt} &= q_{in} - q_{out} \\[2ex] \dfrac{ds}{dt} &= \frac{q_{in}}{v}\left(s_{in} - s\right) - \frac{\rho(.)}{Y(\cdot)} \\[2ex] \dfrac{dx}{dt} &= \rho(.) - \frac{q_{in}}{v}x \end{cases} \qquad [1.7]$$

Let us now define $\rho(.) = \mu(.)x$ where μ is called "specific growth velocity" (we will return to this point in the next section). This assumption is reasonable: having defined the biological reaction as being catalyzed by the presence of biomass, it simply guarantees that the growth velocity is zero in the absence of biomass.

This model is a means to obtain the equations of dynamics of s and x for all three modes of operation of interest that we have outlined in section 1.1.2, namely:

– the closed mode (or *batch*) where $q_{in} = q_{out} = Q_{in} = Q_{out} = 0$, $v(0) = V$, $x(0) = X_0$ and $s(0) = S_0$:

$$
\begin{cases}
\dfrac{ds}{dt} &= -\dfrac{\mu(.)}{Y(\cdot)}x \\[2mm]
\dfrac{dx}{dt} &= \mu(.)x
\end{cases}
\qquad [1.8]
$$

– the semi-continuous (or *fedbatch* mode) where $q_{out} = Q_{out} = 0$, $v(0) = V_0$, $x(0) = X_0$ and $s(0) = S_0$:

$$
\begin{cases}
\dfrac{dv}{dt} &= q_{in} \\[2mm]
\dfrac{ds}{dt} &= \dfrac{q_{in}}{v}(s_{in} - s) - \dfrac{\mu(.)}{Y(\cdot)}x \\[2mm]
\dfrac{dx}{dt} &= \mu(.)x - \dfrac{q_{in}}{v}x
\end{cases}
\qquad [1.9]
$$

– the continuous mode where $q_{in} = q_{out} \neq 0$, $v(0) = v(t) = V$, $x(0) = X_0$ and $s(0) = S_0$. By denoting $d = \frac{q_{in}}{V}$, which is called the "dilution rate", it yields that:

$$
\begin{cases}
\dfrac{ds}{dt} &= d(s_{in} - s) - \dfrac{\mu(.)}{Y(\cdot)}x \\[2mm]
\dfrac{dx}{dt} &= (\mu(.) - d)x
\end{cases}
\qquad [1.10]
$$

In the form of the system [1.10], we recognize in each of the equations a term related to hydrodynamics – here, transportation terms $-dx$ and $d(s_{in} - s)$, respectively – and a reaction term: $\mu(.)x$ and $-\frac{\mu(.)}{Y(\cdot)}x$, respectively. This model is called the "chemostat model" and it is therefore thereto and to a number of its extensions that we dedicate this book. According to the forms that μ and Y will take, we will see that the dynamics of this system is particularly rich. Please note that we will essentially focus on the study of equations [1.10] in which the substrate feed concentration and the feed rate are fixed ($s_{in} = S_{in}$ and $d = D$).

A first extension of [1.10] consists of considering that the dilution rate which applies to the "substrate" compartment is not the same as the dilution rate of the "biomass" compartment. This modeling assumption is useful for accounting for the presence of a biomass retention device: a fraction of this latter remaining confined inside the reactor, the term $-d$ in the dynamics of x of equation [1.10] becomes $-\alpha d$. We will see that this modification causes major complications in the qualitative analysis of the system because it prevents using a common change of variable $z = Ys + x$ (which accounts for the conservation of matter inside the system: at steady-state, part of the available resource s_{in} is transformed into x and the complement is a residual substrate s).

At this stage, one may argue, although we have presented several distinct operating modes, why spend most of our efforts in the analysis of continuous models and not also in addressing *batch* and *fedbatch* systems? In fact, the principal advantages of the continuous operation were at the basis of the device itself, namely our ability, at equilibrium, to "drive" the net growth rate of a microorganism. We will return in more detail to this point in Chapter 2. This reason thus leads us naturally to the question of modeling these growth rates, which we will call hereafter "biological kinetic".

1.2.4. *Biological kinetics*

In the previous sections, we have presented a model describing the dynamics according to which matter is transformed by biological means. However, we have not discussed the way in which the velocity – or kinetics – of the reaction can be modeled. The only assumption that has been made so far is that velocity ρ was written $\rho = \mu(.)x$. If we take the equations of growth in *batch*, we have $\frac{dx}{dt} = \mu(.)x$. By definition, the "specific growth rate" refers to the quantity $\frac{1}{x}\frac{dx}{dt} = \mu(.)$. Monod, in his works on the growth of microorganisms, discovered that the function $\mu()$, with fixed environmental conditions, only depended on the concentration of the limiting substrate. In particular, he introduced the function $\mu(s) = \mu_{max}\frac{s}{s+K_S}$ [MON 50]. This function is zero for $s = 0$ and tends toward μ_{max} when the substrate concentration becomes "large" compared to K_S. In addition, he named K_S the semi-saturation constant, denoting that $\mu(K_S) = \frac{\mu_{max}}{2}$. When $\mu = \mu(s)$, the growth rate is proportional to the population density: moreover, all things being equal to a biomass density twice greater, the concentration of microorganisms grows twice faster. Nonetheless, this expression has given rise to numerous discussions. The first criticism that can be made of this expression is that it does not originate from a law (even if it is sometimes presented and named as such) but from a heuristic approach enabling a two-variable function to replicate data in a satisfactory way. In reality, it is inspired by the Michaelis–Menten expression established in 1913 and that which describes the kinetics of a reaction catalyzed by an enzyme (see [JOH 11]). The latter, even if it is not sufficient to describe complex situations, is based on mechanistic bases since a purely chemical formalism makes it possible to establish it. In the case of microbial growth, the situation is different because it is the product of a very large number of intracellular reactions whose result is observed at the population level. In more complex situations, the use of expressions involving a larger number of parameters is necessary. We will here omit the hundreds of expressions of the literature that have been proposed since Monod's time (see the appendix of Bastin and Dochain's book that lists many expressions of kinetics proposed in the literature [BAS 90]) and we will merely identify the refinements that followed in microbial kinetics modeling:

– it is essential to note here that in most cases, growth rate basically depends on a large number of parameters. For example, it is absolutely intuitive to consider that temperature and pH will play crucial roles in microbial growth: maintaining our refrigerator at $4°$ makes it possible to limit the development velocity of microorganisms and as a result to preserve our food longer. In order to concentrate on the role of one or more limiting substrates, we will consider here situations where these environmental variables will be, if not optimal, at least constant with values somehow "ideal" for microorganisms;

– as it is very often the case, if a substrate is limiting at low concentration, it can also prove toxic when its concentration becomes significant in the medium. In 1968 [AND 68], Andrews has suggested an expression for $\mu(s)$ involving three parameters to describe the growth rate of a microorganism limited by a low concentration of substrate but inhibited when the concentration becomes significant [HAL 30]. This function is written $\mu(s) = \mu_0 \frac{s}{s + K_S + \frac{s^2}{K_I}}$ where μ_0 and K_S are, respectively, the maximum growth rate and the semi-saturation constant in the absence of inhibition and K_I the inhibition constant. We will see that the choice of this kinetics leads to very significant changes in the qualitative properties of biological models. It is particularly interesting to note here that this function actually describes an indirect inhibition phenomenon. Similarly to the Monod equation, this equation is not a law of nature but expresses the fact that at high concentrations, the complex mechanisms involved in fact cause a change of pH and that it is this variation of pH which has, *in fine*, a consequence on growth. The Haldane function[2] thus perfectly illustrates the manner in which very complex phenomena can be reduced to simple modeling by adopting a "macroscopic" point of view – at the population level. The fact remains that if we adopt this very integrating vision, the range of validity of the developed model is limited and that it may prove necessary to write more complicated models capable, for example, of explicitly describing the relationship between biological growth and pH variation in the reactors. This reason has driven us to add at the end of this chapter a section called "Towards a little more realism" in which we give elements for the modeling in order to approach models closer to reality;

– if the previous kinetics do only involve the substrate concentration – they will be known as "substrate-" or "resource-" dependent; some authors have pointed out that the experimental data obtained from complex microbial ecosystems (typically in

---

2 It should be noted that this is the designation which is adopted in bioengineering, although it was Andrews who, inspired by the function proposed by Haldane to describe enzymatic phenomena, suggested the form that we have presented for biological reactions inhibited by substrate.

"mixed cultures") were better reproduced by making use of kinetics depending not only on substrate density, but also on biomass density. This is the reason why, in this case, this will be referred to as of "density-dependent" kinetics. The so-called Contois function is thus written $\mu(s,x) = \mu_{max}\frac{s}{s+K_S x}$. Its particularity is that it decreases when the biomass concentration increases. This function is also called "ratio-dependent" because it is rewritten in the form $\mu(\frac{s}{x}) = \mu_{max}\frac{\frac{s}{x}}{\frac{s}{x}+K_S}$. Similarly to the previous case of inhibition, this function is very helpful when it comes to account for complex aggregation phenomena such as biomass structuring in flocs. In effect, in this case, the part of biomass that is located in the core of the flocs receives the substrate by diffusion only, hence a strong limitation of growth: it follows that at a given microbial population, the overall growth is lower in comparison to a situation in which biomass would not be structured into flocs (see [LOB 06a, HAE 07] or still [HAR 07]). It is interesting to note that almost all the growth functions have long been studied in distinct disciplinary fields (typically in general ecology and microbiology) until these fields converge today, especially in microbial ecology, into a discipline that emerged in the early 2000s (see [JOS 00]). Again, we will see that the properties of the models are modified in a very significant manner when these kinetics are utilized.

1.2.5. *The benefits of the chemostat*

We recall here that the chemostat is originally an experimental device used to study microbial growth and possesses the principal characteristics of continuously operating and ensuring a homogeneous growth medium. The microbial ecosystem — whether implementing a pure or mixed culture — is the functional core of any biological reactor. In the presence of fixed environmental conditions and provided a structure of adequate growth function and a fine identification of its parameters, the publications available in the literature show that the model of the chemostat reproduces the experimental reality particularly accurately by considering all or part of the present ecosystem as functional populations (see [GRA 72] or [GRA 75]). Therefore, it appeared legitimate to make use of it as a "building block", so to speak, in order to study more complex situations, particularly related to the dynamics of diversity. As a result, as early as the 1960s, shortly after its invention, the device has been modernized and has become a prominent tool in microbiology laboratories to characterize all kinds of microbes. However, the "model of the chemostat" as a mathematical object was seen as an "formal entity", exposed in the field of general ecology to theoretically study the properties of trophic chains, questions related to microbial interactions or even to the genesis and the maintenance of diversity. It enjoyed his moment of glory in the 1980s with Hansen and Hubbell's works dedicated to "competitive exclusion", see [HAN 80] — to which we will return in detail later on (see Chapter 3) — before the growth in molecular biology that overshadowed it for a while in biology laboratories. Technically well understood nowadays by engineers and laboratory technicians, it has gained a renewed interest since the 2000s because it satisfies a number of formalization requirements expressed in the field of microbial ecology (see [HOS 05]), to such an extent that microbial ecosystems have today become models for general ecology (see [JES 04]). Moreover, it is still the subject of studies as a mathematical object to the point that we can now say that it constitutes a branch of applied mathematics, certainly modest, but very active. Furthermore, it proposes a formal framework called "theory of the chemostat" centered around a small well-identified community of mathematicians. Finally, the evidence presented above, borrowing the formalism of process engineering, shows that it can actually be considered to constitute a relevant "building block" to move beyond the pure simulation approach and make it possible to address the study of models of complex biological processes in a systematic and generic manner.

The next section presents the essential elements for the understanding of more comprehensive models used in process engineering, notably when it is desirable to model the pH and gas-liquid exchanges. In addition, it specifies a number of experimental constraints useful to know to somehow measure the distance that may exist between the model of the chemostat and the experimental device that it models. This section is not necessary for readers solely interested in the chemostat as a mathematical object. Therefore, they may now skip to Chapter 2.

1.3. Toward "a little more" realism

1.3.1. *Extensions*

The systematic approach that has just been presented (description of the reaction scheme and the continuous dynamic equations of biological reactions) can be expanded to describe more complex biological systems, namely to taking into consideration a scheme involving more reactions, species and/or processes. Consequently, in water treatment by biological means, it is not uncommon to see models requiring dozens of variables paired with one another *via* a scheme implicating biological reactions simultaneously taking place in cascade and/or in parallel. Because expressing this type of system as equations can prove to be very tedious, it is recommended in such situations to use the so-called Gujer matrices.

The Gujer matrix merely describes the reactive part of the biological dynamics and as such transport terms do not appear therein. It is referred to as a matrix, but is actually presented in the form of a table in which can be found:

– in the first column, the set of processes that have to be modeled;

– on the first row, the set of all states (or components) of the system;

– in the last column, the set of kinetics.

Each element of the table (or matrix) contains a yield coefficient. To illustrate this explanation, consider the biological system consisting of the scheme [1.3]–[1.4]. This scheme involves two processes (growth and mortality), three states (s, x and x_d) and two kinetics (one for each process). The Gujer matrix of this system comprises two rows and three columns and is represented in Table 1.1.

Process	x	s	x_d	Kinetics
Growth of x	1	$-\frac{1}{Y(\cdot)}$		$\mu(s)x$
Mortality of x	−1		1	$k_d x$

Table 1.1. *Example of a Gujer matrix for the biological system whose reaction scheme is given by [1.3]–[1.4]*

The representation by means of the Gujer matrix presents several interests. First, it provides a means to test the model "balancing" in the sense where the mass balance, if the variables are all expressed in the same unit, should be able to "be balanced" even if this means adding matter compartments. This closure is due to the fact that what disappears from a compartment must necessarily appear in another compartment. Each row represents a process, balancing is simply tested by verifying that the sum of the components of a row is indeed zero. In the example below, this is the case only for the mortality process but not for growth: it is effectively verified that $1 - \frac{1}{Y(\cdot)} \neq 0$ (unless $Y(\cdot)$ is equal to one which is excluded for the reasons that we already have exposed in previous discussions). In fact, this is explained by the fact that we have included in the model only the result of the part of s which was used for growth without considering its complement, typically used for the creation of the energy required for this growth. Without getting into complicated metabolic details, this model could thus be completed by adding a CO_2 compartment whose excretion by the cell is representative of this energy produced. The new matrix would then assume the form represented in Table 1.2.

Process	x	s	x_d	CO_2	Kinetics
Growth of x	1	$-\frac{1}{Y(\cdot)}$		$\frac{1-Y(\cdot)}{Y(\cdot)}$	$\mu(s)x$
Mortality of x	−1		1		$k_d x$

Table 1.2. *Example of a Gujer matrix for the biological system whose reaction scheme is given by [1.3]–[1.4] with closure of matter in the growth term*

Another advantage of this matrix notation is that it will be possible to deduce the dynamics of the variables from a mere reading of the Gujer matrix. In effect, let us remember that the dynamics of a biological variable is the sum of two types of terms: one related to the modeling of the hydrodynamics (in the case of the chemostat, it designates a transportation term in $D(x_{in}^i - x_i)$) and a reaction term accounting for the formation and disappearance processes of the component under consideration *via* the various processes modeled. The Gujer matrix allows the immediate notation of the latter term in the following manner. For the component i, it is sufficient to obtain the sum of the elements contained in the ith column of the matrix, multiplied, element by element, by the kinetics that can be found in the last column. Let us illustrate this procedure with the example in Table 1.2. The part of the dynamics of x related to growth is given by the sum of the yield coefficient 1 that multiplies the kinetics $\mu(s)x$ (the term associated with growth) added to the mortality term expressed by multiplying the yield coefficient -1 multiplied by the term $k_d x$. Therefrom, it is concluded that $\frac{dx}{dt} = \mu(s)x - k_d x$. In order to obtain the complete equation, the last thing to do is to add the transportation term to yield the dynamic $\frac{dx}{dt} = \mu(s)x - k_d x + D(x_{in} - x)$ or even $\frac{dx}{dt} = \mu(s)x - k_d x - Dx$ if $x_{in} = X_{in} = 0$.

It is important to note that this type of extension — the introduction of new processes in ecosystem models — can be complemented by a more comprehensive description of biological reactions, which takes the form of a scheme where multiple reactions take place simultaneously in parallel and/or cascading, analogously to what is obtained in the case of a trophic chain. Thus, for example, anaerobic digestion, which is the transformation of organic matter into biogas (a mixture of methane and carbon dioxide), is achieved by a microbial consortium involving a very large number of groups of microorganisms within which groups of molecules are transformed into a certain number of products that themselves appear to be the substrate(s) of other groups of microbes. To cite only one example, the AMOCO model[3] or AM2[4], see [BER 01], thus involves two families of molecules combined into two variables, respectively, denoted by s_1 and s_2. It further presupposes a reaction scheme in two stages where a consortium of microorganisms x_1 transforms s_1 into s_2 which in turn is transformed by a consortium x_2 into biogas.

Now that we have presented the main tools to formalize a biological reaction from a perspective that could be qualified as being "macroscopic", the following section is dedicated to the modeling of physicochemical equilibria. Although in the remainder of this book, we do not specifically study this type of model, necessarily more complex than the "simple model of the chemostat", with this complement, we will be able to approximate the description of reality. This will be achieved by establishing a link between the description of the biological phenomena and physicochemical equilibria that govern, in particular, the pH, which is a key variable

---

3 From the name of the European project in which it has been developed.
4 For *Acidogenesis Methanogenesis Model.*

in the functioning of microbial ecosystems. It should be noted that at a fixed pH, we naturally fall back on the chemostat model...

1.3.2. *pH and physicochemical equilibria*

Chemistry tells us that most of the molecules that we immerse in water dissociate in several "species" (here taken within the meaning of chemistry) and that the forms they assume are in equilibrium in the medium depending on the pH. In addition, when it is said that a microorganism needs a substrate, for example nitrogen, this does not specify that it is usually able to assimilate it in ammonium form only and that, in another form, the substrate can be toxic. So far, we have performed balances on the total quantity of elements present in the medium while disregarding the form under which these elements were found. Nonetheless, this form depends on the pH. Thus, for example, nitrogen that is added to water is present in NH_4^+ form (acid, of concentration $[NH_4^+]$) and NH_3 (base, of concentration $[NH_3]$) which form what is called an acid-base pair. In other words, in water, the two following reactions simultaneously occur:

$$\begin{cases} NH_4^+ + H_2O \Rightarrow NH_3 + H_3O^+ \\ NH_3 + H_2O \Rightarrow NH_4^+ + OH^- \end{cases}$$

[1.11]

in which the ratio $\frac{[NH_3]}{[NH_4^+]}$ is related to the pH by the equilibrium formula $pH = pK_a + log(\frac{[NH_3]}{[NH_4^+]})$ where pK_a is a constant (experimentally determined within given environmental conditions; equal to 9.2 in normal conditions of temperature and pressure, which means that when the ratio of the two concentrations is equal, the pH is equal to 9.2).

At a pH greater than 9.2, the nitrogen will appear more in NH_3 form while at a lower pH, nitrogen will be more in NH_4^+ form. However, the modeling of these phenomena can be complicated because of several factors:

– some molecules are in equilibrium in several liquid, aqueous and possibly solid forms (*via* precipitation processes), these equilibria being not only dependent on the pH but also of other environmental variables such as temperature and pressure;

– at lower pH (respectively, larger), some chemical species can again be dissociated and it is then necessary to consider new variables corresponding to these new species if the aim is to correctly model the phenomena being addressed. This new species is itself in equilibrium with the two other forms that were considered until then: it results, therefore, that we are not considering an equilibrium equation but two reactions paired to one another by means of the pH.

Let us illustrate these two situations. First, we recall the case of nitrogen. Previously, we have deliberately omitted to specify that NH_3 is actually a gas whose dissolved part is in equilibrium with its gaseous form. Their respective proportions obey Henry's law which states that $[NH_{3_{aq}}] = K_{NH_3} * P_{NH_3}$, where K_{NH_3} is Henry's constant whose value depends on the gas, on the liquid and on environmental conditions, while P_{NH_3} is the partial pressure of the gas. It is paramount to realize here that the constants that appear in the various equations to which we reference are experimental values obtained for pure bodies. Since they cannot test the infinity of situations that may arise in practice, users will often merely limit themselves to these data and consider that the behavior in mixture is the result of the sum of the behavior of each of the elements if they were considered on their own in the medium. To illustrate the second situation, we will consider the chemistry of inorganic carbon (for details, see [COP 96] from which we extract this example). When dissolved, carbon dioxide is a "diacid", precisely because it disassociates twice, first into hydrogenocarbonate ions HCO_3^- (analogously to what has been written for nitrogen, this first equilibrium is governed by the equilibrium equation [1.12]), and a second time into carbonate ions CO_3^{2-} governed by the equilibrium equation [1.13]. If we denote the total carbon concentration (denoted by TC), we have $TC = [CO_{2,dissolved}] + [HCO_3^-] + [CO_3^{2-}]$ and the concentrations of the different forms of inorganic carbon are related to each other by equations [1.14]:

$$\frac{[HCO_3^-][H^+]}{[CO_{2,dissolved}]} = K_1 \tag{1.12}$$

$$\frac{[CO_3^{2-}][H^+]}{[CO_{2,dissolved}]} = K_2 \tag{1.13}$$

$$\begin{cases} [CO_{2,dissolved}] &= \dfrac{[H^+]^2}{[H^+]^2 + K_1[H^+] + K_1 K_2} TC \\[2ex] [HCO_3^-] &= \dfrac{K_1[H^+]}{[H^+]^2 + K_1[H^+] + K_1 K_2} TC \\[2ex] [CO_3^{2-}] &= \dfrac{K_1 K_2}{[H^+]^2 + K_1[H^+] + K_1 K_2} TC \end{cases} \tag{1.14}$$

the ratios $\dfrac{[CO_{2,dissolved}]}{CT}$, $\dfrac{[HCO_3^-]}{CT}$ and $\dfrac{[CO_3^{2-}]}{CT}$ depending only on the pH.

Having achieved a balance with TC, we have a system of four equations with four unknowns. Supposing that dissociation constants K_1 and K_2 are known, it is therefore possible to calculate the concentrations of the different forms of inorganic carbon and therefrom deduce the medium pH. In practice, in order to model the pH within a biological reaction, assuming that dissociations are very fast compared to biological reactions, algebro-differential models are thus considered in which all the physicochemical equilibriums of existing species are expressed to thereof deduce the concentration of H^+ ions and therefore the pH. It should be noted that these are the considerations that will usually drive engineers to limit the range of validity of the models developed.

1.3.3. *Spatialization*

Previously, we have only mentioned situations in which the enclosure of the reactor was homogeneous. However, it is easily understandable that the more significant the volume of a reactor is going to be, the more this homogeneity condition is likely to be questioned. In addition, if we broaden the "formalization field of the chemostat to the description of natural ecosystems", then there are many situations in which the structuring of the natural space can be seen as more or less large − homogeneous − volumes connecting each others by flows of matter and/or energy. If we have a perfect knowledge of the dynamic behavior of each individual entity, the interest in formalizing this natural space as a network of interconnected chemostats is immediately understood. Note that if we consider flows of matter − whose intensity can be varied − going from a reactor A to a reactor B and *vice-versa*, we are confronted with a situation in which diffusion phenomena can be studied. The first configurations involving interconnected chemostats are known under the name of "gradostat" and have been proposed as early as the 1970s to simulate an environment where gradients of concentration of a limiting substrate can be observed as is often the case in a natural space (see [COO 73] or [LOV 79]). By adopting this approach, it is possible to represent numerous non-homogeneous situations by a network of interconnected reactors. The originality of these approaches is to avoid having to write partial derivative equations, which are more difficult to manipulate than a differential system even if the latter is of large dimension. The properties of the "chemostat" as a basic entity can then be "propagated" in the network, for example by simulations, and the properties emerging of the whole studied at least numerically. It is also interesting to point out that if the biological part is forgotten, these approaches that consist of considering networks of reactors (these are then essentially cascades of reactors in which the output of one is the input of the other, which are studied) have been used in the 1950s to study flows in chemical reactors. In particular, these networks have been addressed precisely to "approximate" the hydrodynamic behavior of non-homogeneous reactors, also called plugflow reactors, which is one of the "ideal" reactors of process engineering (see [CHO 59, CHO 60] or further [CHO 61]). In addition, such configurations have been proposed to bring forward ratio-dependent growths (see [ARD 92]) or still to model biofilm reactors (see [ESC 05]). In all these situations, flows − the flow rates between the different chemostats of the networks under consideration − are *a priori* constant. Already very rich in terms of dynamics, considering the chemostat or a network of chemostats in a context where flows that connect them vary opens very interesting new perspectives as we will see in the next section.

1.3.4. *Recent developments*

At the present moment, the world of microbial ecology is going through a real revolution due to advances in molecular biology. That is how, from the research for the comprehension of evolution and cellular regulation mechanisms, a new disciplinary field called "systems biology" has recently emerged. This branch of science, basically multidisciplinary, seeks to understand the cellular mechanisms at the basis of the functioning of living cells. Its objective is clearly visible: to be able in the coming years to propose an "in silico" cell mimicking in every aspect the functioning of a living cell. This would be achieved by simulating its growth from reading its DNA until its division into two daughter cells. One of the difficulties that confronts researchers is to be able to study these cells within stationary environments, which a chemostat precisely allows. Associated with high-frequency sampling of all the characteristics of the medium but also of the internal cellular metabolites and with the use of all the "omics" measures available, new experimental devices have emerged, based on the principle of the chemostat (homogeneous and operating continuously). They are supposed to be able to provide researchers with the set of data that includes the necessary information for the understanding of studied phenomena. They take advantage not only of the fact that chemostats actually enable equilibria to be obtained, but also that these equilibria can be interacted upon *via* actuators. Operating then in a dynamic mode, they are called "changestats", "accelerostats" or still "decelerostats". One can, for example, distinguish the "turbidostat" (operation with controlled cells concentration), the "A-stat" or the "in-stat" (operations in which a succession of pseudo-equilibriums are obtained before the inflow rate is increased or on the contrary decreased, either continuously or abruptly by means of applying a level), "auxo-stat" (operation with controlled oxygen concentration), etc. In this new context, the conventional chemostat becomes a particular case called "D-stat" (see [ADA 15]).

It seems particularly relevant to make two observations here. The first is that although these devices manifestly appear as carrying a strong application potential, their theoretical study remains to be achieved and may hold a few surprises. For example, Lobry *et al.* have considered a simple situation in which two species are in competition on a limiting resource within a chemostat subjected to a slow variation in the inflow rate. They have established that the model describing this device presented a phenomenon called "stability loss delay". This is a property of dynamical systems that allows the species that finally loses the competition able to sustain themselves "longer than expected in the system" even though the feeding flow rate reaches a value that is unfavorable (see [LOB 09] or [LOB 13]). The second observation is that the devices to which we are referring here are not as recent as it might seem, since a number of them were proposed as early as the 1990s. However, it is rather their utilization in the context of systems biology and, as a result, the renewed interest in the chemostat that is interesting to point out here (see [HOS 05] or still [ZIV 13]).

2

The Growth of a Single Species

In this chapter, we are going to establish the mathematical properties of the model [1.10] which has been described in Chapter 1. The model is:

$$\begin{cases} \dfrac{ds}{dt} & = \; D(S_{in} - s) - \dfrac{\mu(\cdot)}{y(\cdot)}\, x \\[2em] \dfrac{dx}{dt} & = \; (\mu(\cdot) - D)\, x \end{cases} \qquad [2.1]$$

where we are now going to specify the $(\cdot)$ which represents the way in which μ and y depend on the state of the system.

We will start by studying in great detail the simplest possible model that we will call the "minimal model". It is minimal in the sense that if we tried to simplify it slightly more, there will be nothing left of what characterizes a "real" chemostat. For this minimal model, we assume that $s \mapsto \mu(s)$ is a function of the substrate only and that the yield $y(\,) = Y$ is constant. We will perform a very precise mathematical study of this model. It is not very difficult but it is imperative to properly understand all its details because all the subsequent, more complex studies rely on the properties of the minimal model. In particular, the notion of "break-even concentration" is fundamental. In the first section, we establish the mathematical properties that we will interpret in the following section and finally produce some simulations.

In a final section, we will propose four possible extensions to the minimal model. The mathematical treatment will be faster either because it presents no difficulty for the reader who has assimilated the foregoing or, on the contrary, because it is more delicate and falls outside of the context of this book.

2.1. Mathematical properties of the "minimal model"

Throughout this section, $y(.) = Y$ is a constant. The function $s \mapsto \mu(s)$ is continuous and has a continuous derivative, is positive and equal to zero at 0. First, we assume that it is increasing then we introduce the possibility of inhibition phenomena by assuming it is increasing then decreasing.

2.1.1. *General properties*

– *It is always possible to assume that $Y = 1$.* Indeed: we consider the model [2.1]. Let us change the variable:

$$\bar{x} = \frac{x}{Y}$$

and by replacing in the equations it follows that:

$$\left\{ \begin{array}{rcl} \dfrac{ds}{dt} & = & D(S_{in} - s) - \mu(s)\bar{x} \\[2ex] \dfrac{d\bar{x}}{dt} & = & \dfrac{1}{Y}\left(\mu(s) - D\right)Y\,\bar{x} \end{array} \right.$$

which gives:

$$\left\{ \begin{array}{rcl} \dfrac{ds}{dt} & = & D(S_{in} - s) - \mu(s)x \\[2ex] \dfrac{dx}{dt} & = & \left(\mu(s) - D\right)x \end{array} \right. \tag{2.2}$$

It can be seen that Y disappears in the second equation. It suffices to return to the variable x and to set $Y = 1$, which is what we are doing. In other words, up to a multiplicative constant, the variable $\bar{x}$ behaves as x. Since this trick makes it possible to eliminate a parameter – and therefore to simplify the model to be studied – to perform a mathematical study, it is preferable to study [2.2] instead of studying [2.1]. Of course, if we are brought to confront predictions of the model with reality, we will have to reconsider the model [2.1] or to consequently change the units of x.

We call [2.2] the *minimal model*.

– *Existence and uniqueness theorems do apply:* Since it is assumed that μ has a continuous derivative, the second member of [2.2] has continuous partial derivatives and therefore the theorem of existence and uniqueness of solutions can be applied (see theorem A1.1 in Appendix 1).

REMARK 2.1.– Throughout the whole book, except in the appendix dedicated to differential equations (Appendix 1) where specific assumptions will be given, we will always assume, without specifying, that the functions of one or several variables that we introduce are continuous and have derivatives, or partial derivatives, also continuous.

– *The horizontal axis is invariant*: in effect, it is immediately verified that for any $s(0)$ the function:

$$t \mapsto (s(t), x(t)) = \left(S_{in} + (s(0) - S_{in})e^{-Dt}, 0\right)$$

is a solution of [2.2].

– *Solutions remain positive*: Since s and x are quantities (or concentrations) positive or equal to zero, it is important to make sure that solutions with positive initial conditions or equal to zero remain so. For this, it is sufficient to notice that:

$$s = 0 \implies \frac{ds}{dt} = D\,S_{in} > 0$$

and therefore no trajectory can leave the positive orthant $\mathbb{R}^{+2}$ by crossing the vertical semi-axis. Since the horizontal semi-axis is a trajectory and two trajectories cannot cross each other because of the uniqueness of solutions, no trajectory can cross the horizontal semi-axis and leave the positive orthant $\mathbb{R}^2$.

– *An invariant set*: Let $z = s + x$. Let us derive:

$$\frac{dz}{dt} = \frac{ds}{dt} + \frac{dx}{dt} = D(S_{in} - s) - \mu(s)\,x + \mu(s)\,x(t) - D\,x$$

$$\frac{dz}{dt} = D(S_{in} - s) - D\,x = D(S_{in} - (s + x)) = D(S_{in} - z)$$

We can explicitly integrate:

$$\left\{ \begin{array}{l} \dfrac{dz}{dt} = D(Sin - z) \\[2mm] z(0) = s(0) + x(0) \end{array} \right. \implies z(t) = Sin + \left((s(0) + x(0)) - Sin\right)e-Dt \qquad [2.3]$$

One sees that after a transient, we have:

$$s(t) + x(t) \approx S_{in} \qquad [2.4]$$

REMARK 2.2.– The notation $\approx$ is familiar to physicists and engineers; for them, it means that the difference $s(t) + x(t) - S_{in}$ is so small that it is very difficult to measure.

If the initial condition is such that $s(0) + x(0) = S_{in}$, we have "exactly":

$$s(t) + x(t) = S_{in} \qquad [2.5]$$

The segment:

$$I = \{(s, x) : s \geq 0,\ x \geq 0,\ s + x = S_{in}\}$$

is thus an attractive invariant set. In this set, since $s + x = S_{in}$, s can be replaced by $S_{in} - x$ in the second equation of [2.1] which gives the differential equation in x:

$$\left\{ \frac{dx}{dt} = \big(\mu(S_{in} - x) - D\big)x \right. \qquad [2.6]$$

whose behavior is therefore known as soon as the graph of the function $x \mapsto \varphi(x) = (\mu(S_{in} - x) - D)x$ is known. The equality $S_{in} = s + x$ reflects the fact that in this model where the yield is equal to 1 the amount of consumed substrate is transformed into an equal amount of biomass.

– *Solutions are bounded.* This follows immediately from the fact that $t \mapsto z(t)$ is a bounded function, that $s + x = z$ and that s and x are positive or zero.

– *Equilibria.* The equilibria of [2.1] are the solutions of:

$$\left\{ \begin{array}{rcl} 0 & = & D(S_{in} - s) - \mu(s)\,x \\[2mm] 0 & = & (\mu(s) - D)\,x \end{array} \right. \qquad [2.7]$$

There is always the solution $(S_{in}, 0)$ known as the "washout" solution (see section 2.1.4 for the explanations about the term "washout"). The other solutions are:

$$(s^*, x^*)$$

where s^* is a value of s such that $\mu(s^*) = D$ and $x^* = S_{in} - s^*$.

From now on, to continue the discussion on equilibria, we have to specify a few properties of the function μ.

EXERCISE 2.1.– Let $f \mapsto f(x)$ be a continuous and differentiable application of $\mathbb{R}$ in itself and the differential equation:

$$\frac{dx}{dt} = f(x) \qquad [2.8]$$

Let $e_1 < e_2$ be two consecutive zeros of f such that $e_1 < x < e_2 \Rightarrow f(x) > 0$. Show that if $x(t)$ is a solution with initial condition x_o such that $x_o\ e_1 < x_o < e_2$, it is then strictly increasing from e_1 to e_2 for t increasing from $-\infty$ to $+\infty$.

2.1.2. *The function μ is monotonic and bounded*

We assume that the function μ is of the "Monod type", that is:

– μ is defined for $s \geq 0$ and bounded;

– zero for $s = 0$;

– such that $\mu'(s) > 0$ thus strictly increasing.

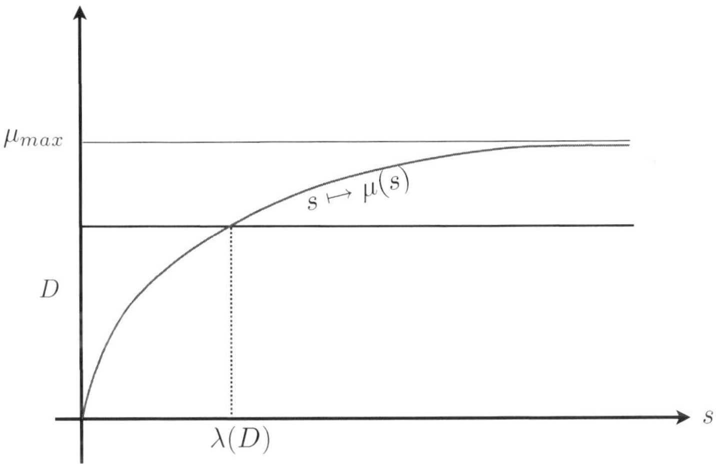

Figure 2.1. *μ function of the "Monod type": we have denoted μ_{max} the upper bound (not reached) ot μ*

NOTATION 2.1.– We have used the notations $\frac{ds}{dt}$ and $\frac{dx}{dt}$ to designate the derivative *with respect to time* of concentrations. For the function $s \mapsto \mu(s)$, we use the notation $\mu'(s)$ to designate the derivative.

In the notation $\frac{ds}{dt} = \cdots - \mu(s)x$; $\frac{ds}{dt} = +\mu(s)x - \cdots$ the term $\mu(s)$ is a harvesting rate in the first equation and a growth rate in the second, which depends exclusively on the value s of the substrate concentration (we will see later how μ could also depend on x); therefore, the larger the substrate concentration is, the more significant the specific growth velocity of microorganisms. It is assumed that there is no inhibition phenomenon (we will see later the possibility of inhibition).

An increasing and bounded function has an upper bound that we denote:

$$\mu_{max} = \sup_{s>0} \mu(s)$$

(the expression "μ_{max}" means "maximum of μ" which is a little bit inappropriate at the mathematical level since the upper bound of a strictly increasing function is never reached).

The Monod function is the function:

$$\mu(s) = \frac{\mu_{max}\, s}{k_s + s} \qquad\qquad [2.9]$$

It is obviously of the "Monod type" and its upper bound is the parameter μ_{max}. The constant k_s is called the "semi-saturation" constant.

2.1.2.1. *Equilibria*

Equilibria are pairs (s_e, x_e) for which the second members of [2.2] are zero; thus $t \mapsto (s_e, x_e)$ is a constant solution of [2.2], and as a result an equilibrium (see Appendix, section A1.1).

– *The "washout" equilibrium.* This is the equilibrium:

$$E_0 = (S_{in}, 0)$$

– *Equilibrium with biomass.* When $D < \mu_{max}$, we denote $\lambda(D)$ the unique s such that $\mu(s) = D$ (it is unique since $s \mapsto \mu(s)$ is strictly increasing) otherwise, we write $\lambda(D) = +\infty$. If $s < \lambda(D)$ the growth velocity of x :

$$\frac{dx}{dt} = (\mu(s) - D)x$$

is strictly negative, whereas if $s > \lambda(D)$ the growth velocity is strictly positive.

DEFINITION 2.1.– *The quantity $\lambda(D)$ is called the break-even concentration.*

This is the first time that we meet this quantity which will play an important role in what follows[1].

Let $D < \mu(S_{in})$; define $s^* = \lambda(D)$ and $x^* = S_{in} - s^*$. Then :

$$E_1 = (s^*, x^*) \qquad\qquad [2.10]$$

---

1 Also known as *seuil de croissance* in French.

is an equilibrium with positive biomass. In effect, on the one hand $s^* = \lambda(D) \implies \frac{dx}{dt} = (\mu(s^*) - D)x^* = 0$, and on the other hand by adding the second two members, we obtain $0 = D(S_{in} - s^* - x^*)$ from which $x^* = S_{in} - s^*$.

If $D > \mu(S_{in})$, we have $s^* = \lambda(D) > S_{in}$ and the equation $0 = D(S_{in} - s^* - x)$ has no solution with $x > 0$ (see Figure 2.2).

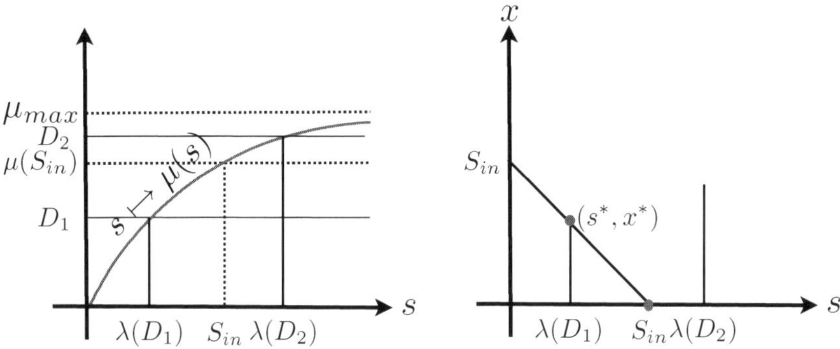

Figure 2.2. *Equilibria: washout equilibrium (red); the equilibrium with positive biomass (blue) exists only for $D < \mu(S_{in})$. For a color version of this figure, see www.iste.co.uk/harmand/chemostat.zip*

2.1.2.2. *Local stability of equilibria*

We know that an equilibrium is locally exponentially stable (see section A1.2) if the real parts of the eigenvalues of the Jacobian matrix at this point are strictly negative. The Jacobian matrix (s, x) [2.2] is:

$$J(s, x) = \begin{bmatrix} -D - \mu'(s)x & -\mu(s) \\ \\ \mu'(s)x & \mu(s) - D \end{bmatrix} \qquad [2.11]$$

– *"Washout" equilibria.* Evaluated at $E_0 = (S_{in}, 0)$ the Jacobian matrix is:

$$J(E_0) = \begin{bmatrix} -D & -\mu(S_{in}) \\ \\ 0 & \mu(S_{in}) - D \end{bmatrix} \qquad [2.12]$$

The two eigenvalues are $-D$ and $\mu(S_{in}) - D$. Therefore, if $D > \mu(S_{in})$ this is a locally exponentially stable (**LES**) equilibrium, if $D < \mu(S_{in})$ this is an unstable equilibrium (a saddle).

REMARK 2.3.– The case $D = \mu(S_{in})$ deserves a comment.

– from a mathematical point of view this is usually called a *bifurcation value*: if we think of D as a parameter, for $D < \mu(s_{in})$ the system has a different behavior than it has for $D > \mu(S_{in})$, LES in the first case, unstable in the second. It is generally more complicated to determine the behavior of the system for a bifurcation value; in our specific case, we see that the Jacobian matrix analysis is insufficient since it has an eigenvalue equal to zero. All we can say is that the system is not LES. In fact, we can show that it is asymptotically stable;

– from practical a point of view, this case can be considered as non-generic.

EXERCISE 2.2.– Demonstrate that when $D = \mu(S_{in})$, the washout equilibrium is stable.

– *Equilibrium with biomass.* For $D < \mu(S_{in})$, the Jacobian matrix $E_1 = (s^*, x^*)$ is:

$$J(E_1) = \begin{bmatrix} -D - \mu'(s^*)(x^*) & -\mu(s^*) \\ \\ \mu'(s^*)(x^*) & (\mu(s^*) - D)x^* \end{bmatrix} \qquad [2.13]$$

but, since $s^* = \lambda(D)$, we have $\mu(s^*) = D$ and the Jacobian matrix is reduced to:

$$J(E_1) = \begin{bmatrix} -D - \mu'(s^*)(x^*) & -D \\ \\ \mu'(s^*)(x^*) & 0 \end{bmatrix} \qquad [2.14]$$

The trace is strictly negative and the determinant positive, therefore the eigenvalues have strictly negative real parts. This equilibrium is locally exponentially stable and since the roots are real there are no oscillations around equilibrium. The expression "locally" means that if initial conditions are sufficiently close to the equilibrium the solutions tend toward the equilibrium; see section A1.1 in Appendix 1 for accurate mathematical definitions. This information is summarized in Table 2.1:

S_{in}	$\lambda(D) < S_{in}$	$\lambda(D) = S_{in}$	$\lambda(D) > S_{in}$
E_0	Unstable	Stable	LES
E_1	LES	Does not exist	Does not exist

Table 2.1. *Local stability of the equilibria of [2.2] for "Monod type"* μ

2.1.2.3. *Global stability*

In the previous section, the use of the result from the Jacobian matrices makes it possible only to conclude for local stabilities, but in fact we have the following proposition which is stronger.

PROPOSITION 2.1.– Let the system be [2.2].

– if $D < \mu(S_{in})$, the washout equilibrium E_1 of [2.2] is globally asymptotically stable (GAS) in the strictly positive orthant;

– if $D \geq \mu(S_{in})$, the "washout" equilibrium E_0 of [2.2] is globally asymptotically stable (GAS) in the positive orthant.

PROOF.– Given the definition of GAS (see section A1.2) and the local stabilities that we have just proved, it is sufficient to show that:

– if $D < \mu(S_{in})$ any solution of initial condition (s_o, x_o) with $x_o > 0$ tends to E_1;

– if $D \geq \mu(S_{in})$ any solution tends toward E_0.

We use the "method of isoclines" described in section A1.4.2 in Appendix 1. The eager reader can skip this demonstration and return later. We consider the case $D < \mu(S_{in})$. We therefore draw the isoclines of [2.2] so as to delimit domains where [2.2] is monotonic. In Figure 2.3a, we have represented:

– in blue, the graph of $x = \frac{D(S_{in}-s)}{\mu(s)}$ (the reader will make sure that it correctly has the indicated shape) which constitutes the isocline $\frac{ds}{dt} = 0$;

– in red, the line $s = \lambda(D)$, part of the isocline $\frac{dx}{dt} = 0$; the other part being the axis $x = 0$;

– the "washout" equilibrium $(S_{in}, 0)$ (the red dot) and the equilibrium with biomass $E_1 = (s^*, x^*)$ (the blue dot);

– an initial condition (s_o, x_o) such that $s_o + x_o > S_{in}$. Let $(s(t), x(t))$ be the corresponding solution. Since the solutions are bounded, there exists s_m such that for $t > 0$ we have $s(t) \leq s_m$ and x_m such that $x(t) \leq x_m$;

– in the four closed bounded domains $\mathcal{D}_i$ $i = 1, 2, 3, 4$ the system [2.2] is monotonic and signs are indicated.

According to proposition A.9, a solution originating from a point of $\mathcal{D}_1$ (pink area) can possibly tend toward E_1 (it is necessary that $x_o < x^*$ since $t \mapsto x(t)$ is increasing) or leave $\mathcal{D}_1$. If it leaves $\mathcal{D}_1$, it must be through $s = \lambda D$ and then it enters $\mathcal{D}_2$ (green area), which it can leave only along the isocline $\frac{dx}{dt} = 0$ to enter $\mathcal{D}_3$ (emerald area) where it can only tend toward E_1 or leave along the isocline $\frac{dy}{dt} = 0$ to enter $\mathcal{D}_4$ (white area) which it can only leave along the isocline $\frac{dx}{dt} = 0$ to enter again $\mathcal{D}_1$ and

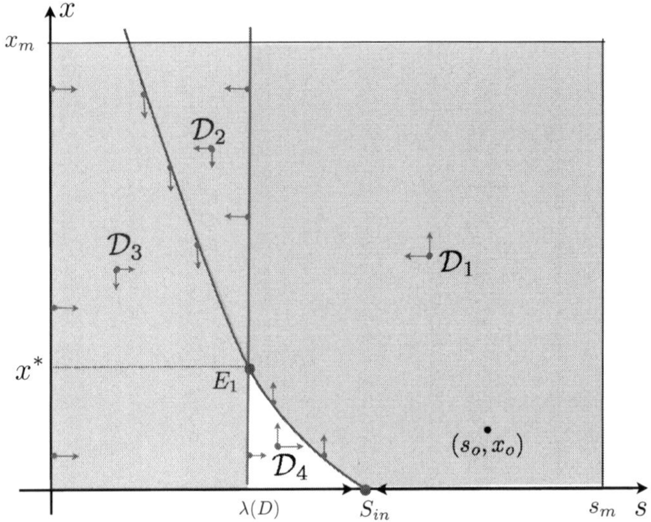

a) The isoclines of [2.2]

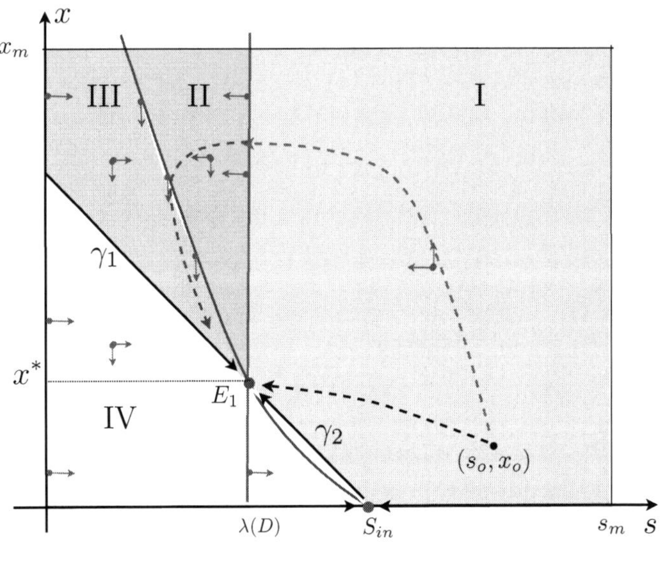

b) Global stability of E_1

Figure 2.3. *Monod model when $D < \mu(S_{in})$: global stability of the equilibrium E_1. For a color version of this figure, see www.iste.co.uk/harmand/chemostat.zip*

so on. *A priori*, there is nothing that prevents the solution from spiraling around E_1 and from becoming trapped in a limit cycle which would prevent it to tend toward E_1. The "method of the isoclines" *on its own does not make it possible to conclude on this point.*

However, in reality solutions cannot loop around E_1 for the following reason. In Figure 2.3(b), we have added to Figure 2.3(a), the invariant set:

$$I = \{(s, x) : s \geq 0,\ x \geq 0,\ s + x = S_{in}\} = \gamma_1 \cup E_1 \cup \gamma_2$$

that divides the space into two regions that cannot communicate with each other. Here, γ_1 and γ_2 are the two trajectories of the invariant set that tend toward the equilibrium E_1. Let us call I, II and III the sections of the domains $\mathcal{D}_1$, $\mathcal{D}_2$ and $\mathcal{D}_3$ located above I. Recalling the description of the movement of the resulting trajectory of (x_o, y_o), we represent in dotted lines two *a priori* different possibilities (naturally, only one of the two is actually achieved since there is uniqueness of solutions):

– if the trajectory resulting from (s_o, x_o) remains inside of I, it tends toward an equilibrium that may only be E_1 since $x(t)$ is increasing. If the path leaves I, it enters II;

– a trajectory originating from a point of II can only enter III;

– since the field [2.2] is entering in III, a solution originating from III can only tend toward E_1 which is the unique equilibrium of [2.2] belonging to III.

We have therefore shown that any trajectory coming out from a point located above the invariant set I tends toward E_1. Using the same kind of arguments, we see that a trajectory coming out of IV also tends toward E_1.

Readers are also required to make sure by themselves that when $\mu(S_{in}) < D$ all solutions tend toward the washout equilibrium. ■

REMARK 2.4.– When $D > \mu(S_{in})$, it is possible to directly ensure that all solutions tend toward the washout equilibrium without the need for geometrical analysis. In effect, in this case, since μ is strictly increasing, there exist $a > 0$ and $\delta > 0$ such that $\mu(S_{in} + a) \leq D - \delta$; since $s + x \to S_{in}$ for t large enough, we have $s(t) < S_{(in)} + a$ and thus $\frac{dx(t)}{dt} \leq -\delta x(t)$. More informally, if $D > \mu(S_{in})$, for t large enough, s is always smaller than the growth threshold and biomass can only decrease toward 0.

Therefore, we can clarify Table 2.1 with Table 2.2.

2.1.3. *The function μ is not monotonic*

In the previous section, there was a unique s such that $\mu(s) = D$. This does no longer hold when μ is not strictly monotonic. We are going to detail the case where μ is known as being of the "Haldane type", that is:

S_{in}	$\lambda(D) < S_{in}$	$\lambda(D) = S_{in}$	$\lambda(D) > S_{in}$
E_0	Unstable	GAS	GAS
E_1	GAS	Does not exist	Does not exist

Table 2.2. *Local stability of the equilibria of [2.2] for "Monod type"* μ

– defined for $s \geq 0$ as positive and zero for $s = 0$;

– there exists $s_m > 0$ such that $s \in [0, s_m[\Rightarrow \mu'(s) > 0$ and $s \in]s_m, +\infty) \Rightarrow \mu'(s) < 0$;

– $\lim_{s \to +\infty} s(t) = 0$.

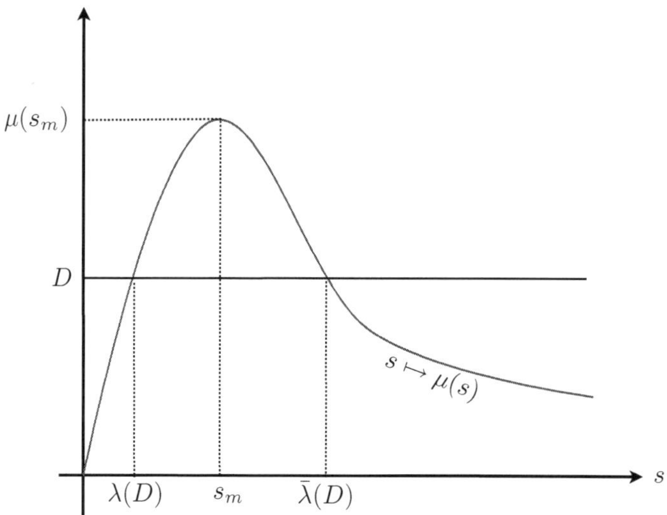

Figure 2.4. *Solutions of the equation* $\mu(s) = D$

Such a function is obviously first strictly increasing, it reaches its maximum $\mu(s_m)$ in s_m then is strictly decreasing. *Haldane*'s kinetics (see Chapter 1) which are:

$$\mu(s) = \mu_0 \frac{s}{s + K_S + s^2/K_I} \tag{2.15}$$

are of course of the "Haldane type". For this type of function, when $D < \mu(s_m)$, the equation $\mu(s) = D$ has two solutions that we denote $\lambda(D)$ and $\bar{\lambda}(D)$ with $\lambda(D) < \bar{\lambda}(D)$ (see Figure 2.4).

2.1.3.1. *Equilibria*

As in previous cases, equilibria are the pairs (s_e, x_e) for which the second members of [2.2] are zero.

– *The "washout" equilibrium.* For any D, we have the equilibrium:

$$E_0 = (S_{in} \ ; \ 0)$$

– *Equilibria with positive biomass.* If $\mu(\lambda(D)) < S_{in}$, we have the equilibrium:

$$E_1 = (s^*, x^*)$$

with $s^* = \lambda(D) \quad x^* = S_{in} - \lambda(D)$. If $\mu(\bar{\lambda}(D)) < S_{in}$, we have in addition the equilibrium:

$$\bar{E}_1 = (\bar{s}^*, \bar{x}^*)$$

with $\bar{s}* = \bar{\lambda}(D) \quad \bar{x}^* = S_{in} - \bar{\lambda}(D)$.

2.1.3.2. *Local stability of equilibria*

The Jacobian matrix is the same as in the case when μ is monotonic. For the three equilibria, it yields:

$$J(E_0) = \begin{bmatrix} -D & -\mu(S_{in}) \\ 0 & \mu(S_{in}) - D \end{bmatrix} \tag{2.16}$$

$$J(E_1) = \begin{bmatrix} -D - \mu'(s^*)x^* & -D \\ \mu'(s^*)x^* & 0 \end{bmatrix} \tag{2.17}$$

$$J(\bar{E}_1) = \begin{bmatrix} -D - \mu'(\bar{s}^*)\bar{x}^* & -D \\ \mu'(\bar{s}^*)\bar{x}^* & 0 \end{bmatrix} \tag{2.18}$$

What differentiates $J(E_1)$ from $J(\bar{E}_1)$ is the fact that in the first case we have $\mu'(s^*) > 0$ and in the second case $\mu'(\bar{s}^*) < 0$ which changes the nature of the eigenvalues. We ask the reader to verify the accuracy of Table 2.3. It can be seen that when $\bar{\lambda}(D) < S_{in}$, there are two exponentially stable equilibria, the "washout" E_0 and E_1; the two basins of attraction of the two stable equilibria are separated by the two trajectories that reach the saddle E_2; one says that there is *bistability*.

S_{in}	$S_{in} < \lambda(D)$	$\lambda(D) < S_{in}) < \bar{\lambda}(D)$	$\bar{\lambda}(D) < S_{in}$
E_0	LES	Unstable	LES
E_1	Does not exist	LES	LES
E_2	Does not exist	Does not exist	Unstable

Table 2.3. *Equilibriums of 2.2 for "Haldane type"* μ

EXERCISE 2.3.– We leave it to the reader to establish in Table 2.3 if equilibria are stable locally or globally and to study the case $\lambda(D) = S_{in}$ and $\bar{\lambda}(D) = S_{in}$. Find these results after the change of the variable $z = s + x$.

2.1.4. *Interpretations*

2.1.4.1. *Washout*

From the previous mathematical study, it emerges that the model [2.1], whether μ be of the Monod or Haldane type, has always the equilibrium:

$$E_0 = (S_{in}, 0)$$

which corresponds to a reactor without any biomass. This is referred to as washout equilibrium (or also *washed out*). According to the value of D, this equilibrium may be stable or unstable. The mathematical model tells us that absolute "washout" is never attained *in finite time*. Indeed, equilibria being solutions, the uniqueness of the solutions forbids that a solution originating from a point other than the equilibrium may reach it in finite time. Therefore, in case of "washout", the *biomass $x(t)$* will *tend toward* 0 when t tends to infinity. Nonetheless, from a practical point of view the model no longer represents anything when $x(t)$ is smaller than a certain threshold which depends on the real system modeled. This does not mean that below this threshold there is no more biomass but that the model can no longer predict anything reasonable.

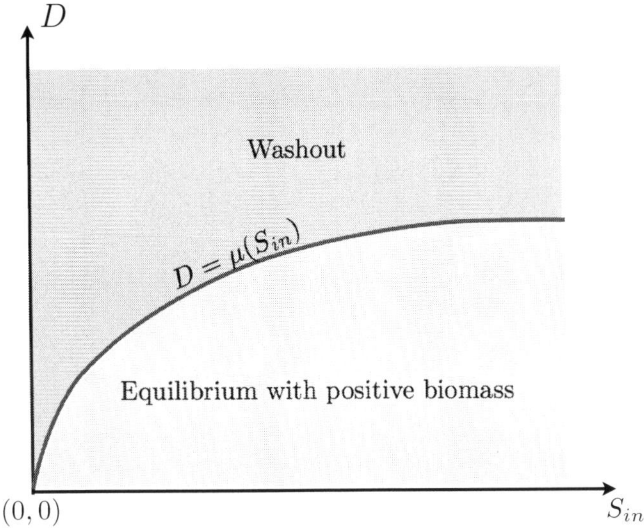

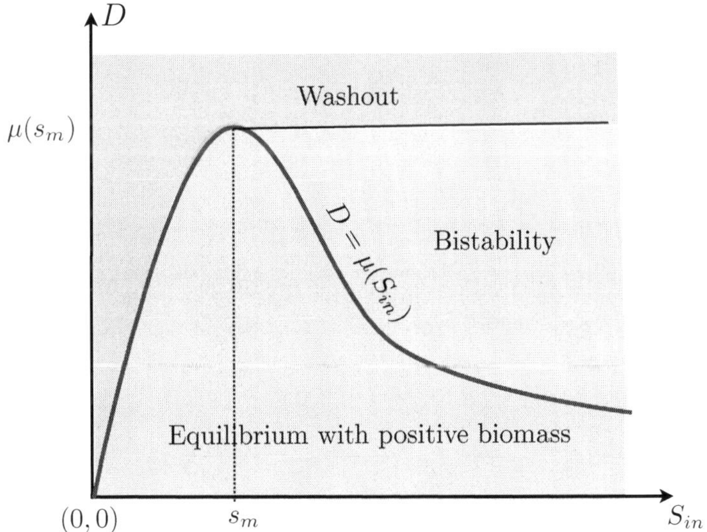

Figure 2.5. *Operational diagram. On the left, "Monod type" μ, on the right "Haldane type". For a color version of this figure, see www.iste.co.uk/harmand/chemostat.zip*

2.1.4.2. *Strong densities*

Since the function $s \to \mu(s)$ is continuous and equal to zero at 0, when D tends toward 0, the break-even concentration $\lambda(D)$ also tends toward 0 and $S_{in} - \lambda(D)$ tends to S_{in}. If S_{in} is significant, the density of the cells of the microorganisms can be so important that the assumptions for the model to be valid will no longer be satisfied.

2.1.4.3. *The operating diagram*

When picturing a laboratory reactor, besides physical-chemical parameters such as pH or temperature, there are two essential parameters that can be manipulated which are the flow rate D on the one hand, and in-flow concentration S_{in} on the other: these are what we called "inputs" in Chapter 1. We have seen that the asymptotic behavior of the chemostat depends on the relations between $\lambda(D)$ and S_{in} which are summarized in Tables 2.2 and 2.3. A slightly more accurate way of representing things is to localize the different possible outcomes in the space of the parameters (S_{in}, D), when S_{in} and D are fixed, with respect to the graph of the function $S_{in} \mapsto \mu(S_{in})$. This is what is called the *operating diagram*. Figure 2.5 represents the operating diagrams of the cases where μ is of the "Monod type" then of "Haldane type".

2.2. Simulations

In this section, we illustrate the theoretical results of the previous section by means of simulations.

When the available system is explicit, it is possible to simulate with a computer the solutions of the differential system. Firstly, the computer is required to calculate solutions (for example, the user could himself program the Euler scheme (see section A1.4) or by making use of software "solvers" such as Matlab or Scilab) and then by displaying the result. In the case of dimension 2 of interest to us here, there are two possible modes of representation:

– either displaying with axes (os, ox) the sequences (s_k, x_k) produced by the computer and the trajectories (or orbits) are obtained (see section A1.1.1) in the phase space;

– or displaying with axes (ot, os) and (ot, ox) the sequences (t_k, s_k) and (t_k, x_k) produced by the computer and thus obtaining the graph of the solutions $t \mapsto s(t)$ and $t \mapsto x(t)$. This method gives information on transients that can be useful because the representation in the phase space does not provide information on the velocity with which the trajectories are followed.

It should be noted that in most cases the integration step dt is so small that (t_k, s_k) and (t_{k+1}, s_{k+1}) always belong to the same pixel or to two neighboring pixels which gives the impression of a continuous curve.

2.2.1. *Simulations in the phase space*

2.2.1.1. *The function μ is a Monod function (Figure 2.6)*

The model [2.2] is simulated with the Monod function as function μ:

$$\mu(s) = \frac{s}{0.2 + s}$$

therefore the system is:

$$\begin{cases} \dfrac{ds}{dt} &= D(S_{in} - s) - \dfrac{s}{0.2 + s}\, x \\[2mm] \dfrac{dx}{dt} &= \left(\dfrac{s}{0.2 + s} - D \right) x \end{cases} \qquad [2.19]$$

The simulations are performed with $S_{in} = 2$; in this case, we have $\mu(S_{in}) = 0.90909...$

– simulation performed with $D = 0.8$ which is thus much smaller than $\mu(S_{in})$; the equilibrium is $(0.8, 1.2)$. The saddle $(S_{in}, 0)$ can be observed;

– simulation performed with $D = 0.92$ which is thus larger than $\mu(S_{in})$; the globally stable washout equilibrium $(S_{in}, 0)$ can be observed.

2.2.1.2. *The function μ is a Haldane (Figures 2.7 and 2.8)*

The model [2.2] is simulated with the Haldane function as function μ:

$$\mu(s) = \frac{m\,s}{K + s + s^2/I} \qquad [2.20]$$

therefore the system is:

$$\begin{cases} \dfrac{ds}{dt} &= D(S_{in} - s) - \dfrac{m\,s}{K + s + s^2/I}\, x \\[2mm] \dfrac{dx}{dt} &= \left(\dfrac{m\,s}{K + s + s^2/I} - D \right) x \end{cases} \qquad [2.21]$$

In all the simulations of Figures 2.7 and 2.8, we have set the parameters:

$$m = 5; \quad I = 0.2; \quad K = 0.5$$

and we have successively taken $D = 1$, $D = 0.6$, $D = 0.5$ and finally $D = 0.4$. In addition to the phase portrait, we have drawn the graph of μ and a leader line that determines $s^* = \lambda(D)$ and $\bar{s}^* = \bar{\lambda}(D)$.

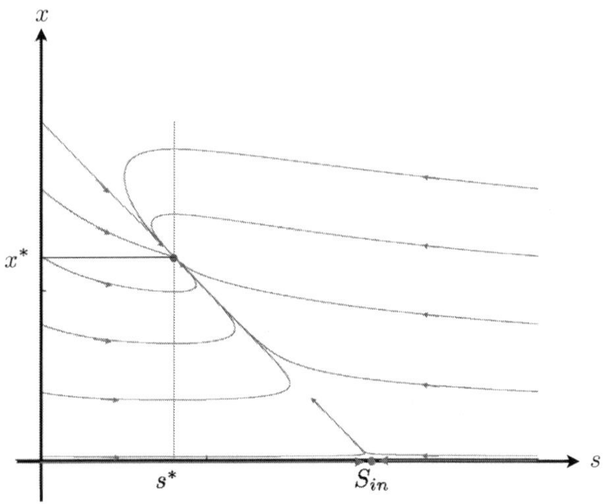

a) Simulation of [2.19]: $D < \mu(S_{in})$

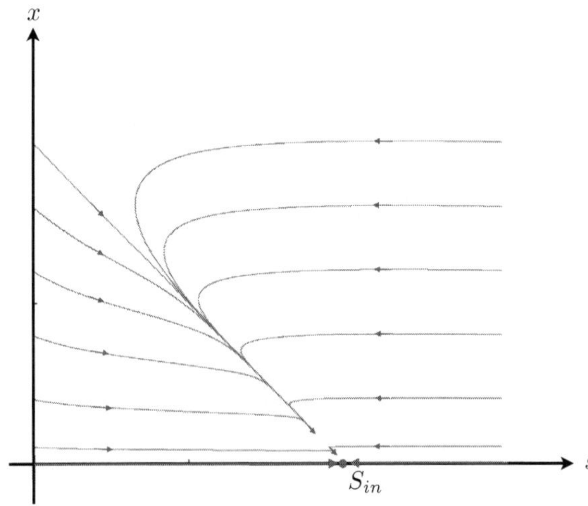

b) Simulation of [2.19]: $D > \mu(S_{in})$

Figure 2.6. *Phase portrait: Monod model. For a color version of this figure, see www.iste.co.uk/harmand/chemostat.zip*

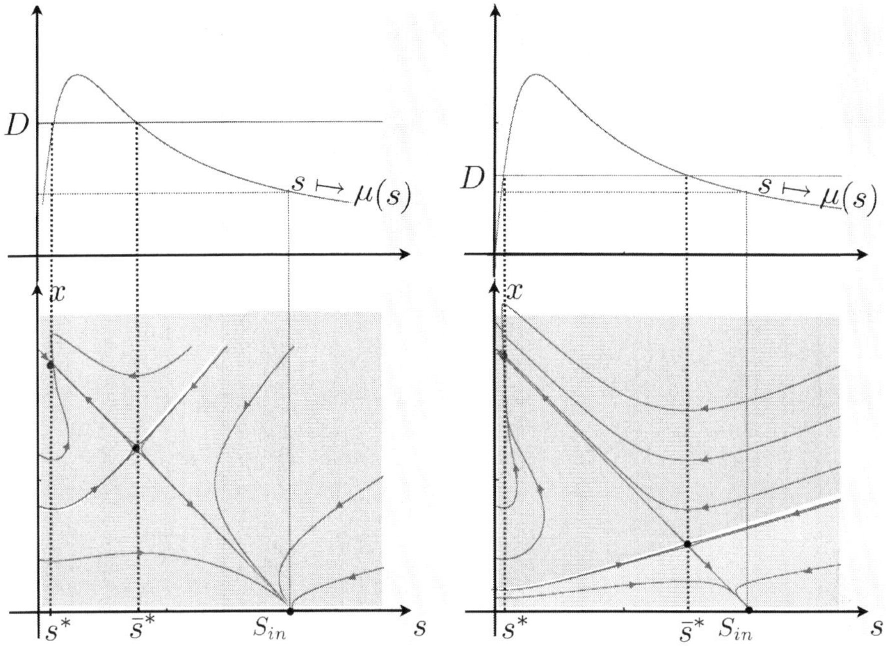

Figure 2.7. *Phase portrait of [2.21]: $D = 1$ on the left;
$D = 0.6$ on the right. For a color version of this figure, see
www.iste.co.uk/harmand/chemostat.zip*

It can be observed that the basin of attraction (see section A1.2) of the washout (in pink) decreases for the benefit of the basin of attraction of the equilibrium with biomass to completely disappear for $D = 0.4$.

2.2.2. *Transients*

In Figures 2.9 and 2.10, 8 simulations of s and x *with respect to time* of [2.19] can be observed (therefore the Monod case), all having $s = 0$ and $x = 0.2$ as initial conditions:

$$(s(0),\ x(0)) = (0,\ 0.2)$$

The duration of the integration is 50 units of time and $S_{in} = 2$. The only parameter that changes is D that increases from $D = 0.5$ to $D = 1.4$. We thus pass through the "washout" value $D = 0.90909\cdots$. Since $s(0) = 0$, it can be observed that it starts with a decrease in x then an increase toward its equilibrium value when the latter is greater than $x(0)$.

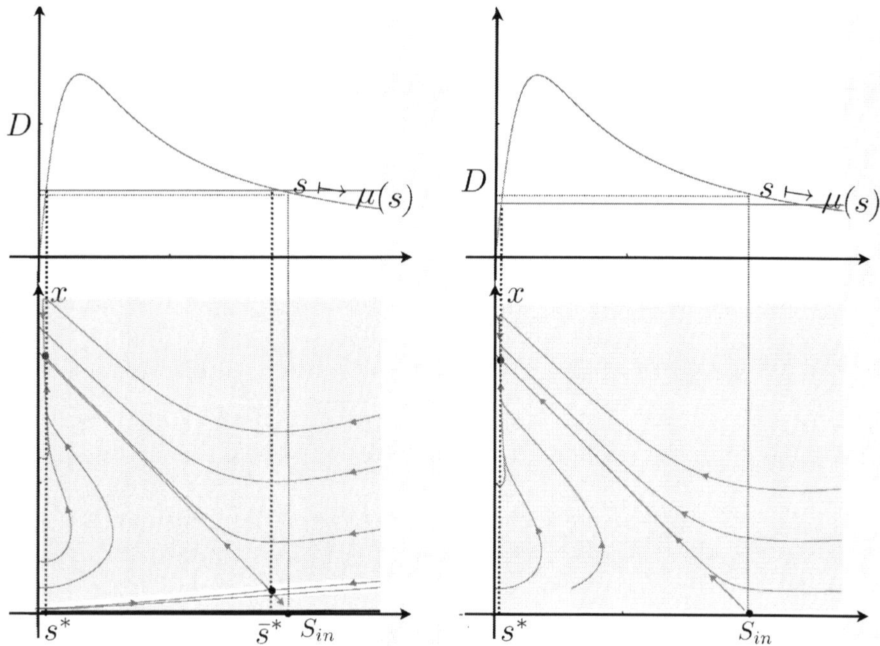

Figure 2.8. *Phase portrait of [2.21]: $D = 0.5$ on the left; $D = 0.4$ on the right. For a color version of this figure, see www.iste.co.uk/harmand/chemostat.zip*

COMMENTS ON FIGURE 2.9.– We have the following table of equilibria values according to D. The values are directly calculated from the model.

D	0.500	0.800	0.820	0.840
s^*	0.200	0.800	0.911	1.050
x^*	1.800	1.200	1.089	0.950

For $D = 0.5$, the transient is fairly brief, in the order of 10 units of time. The duration of the transient increases with D; for $D = 0.84$, the equilibrium is not reached after 50 time units.

COMMENTS ON FIGURE 2.10.– In this figure, the first two values of D are less than the "washout value" ($D_l = 0.90909\cdots$) and close enough. The other two are significantly larger.

D	0.880	0.905	1.912	1.400
s^*	1.466	1.905	2	2
x^*	0.534	0.095	0	0

It can be observed that transients become extremely long when approaching (0.905 and 0.912) the "washout" value; they break down into a relatively short transient where $s(t)$ is increasing rapidly toward a value close to $S_{in} - x(0)$, and then s and x tend very slowly toward their equilibrium value.

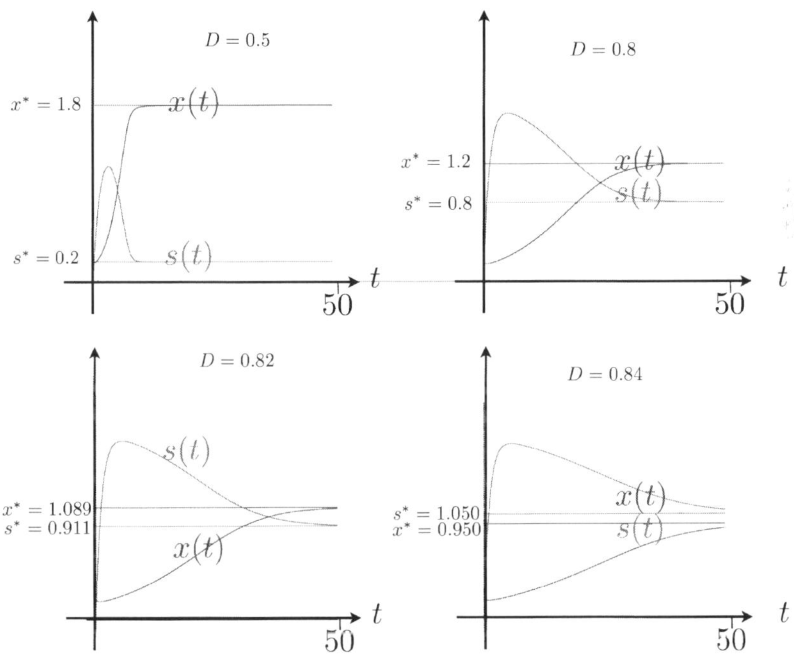

Figure 2.9. *Simulation of [2.19]: comments in the text. For a color version of this figure, see www.iste.co.uk/harmand/chemostat.zip*

2.3. Some extensions of the minimal model

As indicated in the title of the section, we are going to proceed with a few extensions to the minimal model. We successively examine:

– the case in which there is biomass in the feed;

– the case in which the out-flow D_x of the chemostat is not identical to the out-flow D;

– the case in which the function $\mu(\cdot)$ depends not only on s but also on x;

– the case in which the yield Y is not assumed to be constant but may depend on the amount of substrate.

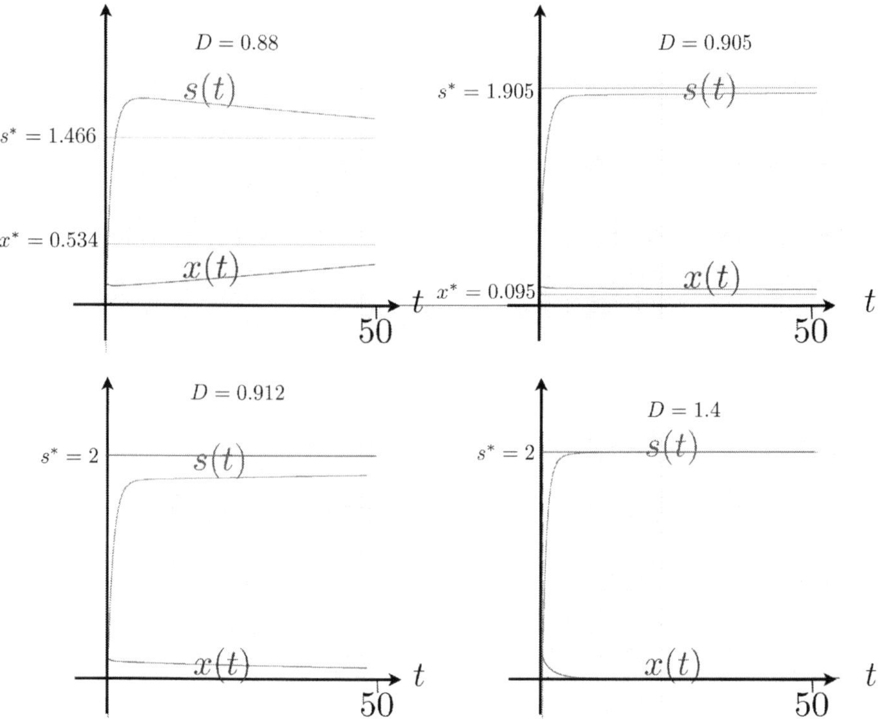

Figure 2.10. *Simulation of [2.19] : comments in the text. For a color version of this figure, see www.iste.co.uk/harmand/chemostat.zip*

Some extensions are mathematically easy, others are less. This will be an opportunity for us to introduce examples of important mathematical techniques (Liapunov functions, Poincaré theorem, Bendixson - Dulac criterion) which complement the isocline analysis and are explained in the appendices.

2.3.1. *Presence of biomass in the feed*

In numerous applications, biomass is already present in the liquid coming into the chemostat. This is the case, for example, for bioreactors for the treatment of wastewater where waters to be treated obviously contain all kinds of bacteria. In the presence of a concentration $X_{in} > 0$ in the in-flow, the minimal model [2.2] becomes:

$$\left\{ \begin{array}{rcl} \dfrac{ds}{dt} &=& D(S_{in} - s) - \mu(s)\,x \\[2ex] \dfrac{dx}{dt} &=& D(X_{in} - x) + \mu(s)x \end{array} \right. \qquad [2.22]$$

The orthant $(\mathbb{R}^+)^2$ remains invariant but the horizontal axis is no longer so, as if it were in the minimal model.

2.3.1.1. *Evolution of the total concentration*

We have the same conservation property as for the minimal model: if we set

$$z = s + x$$

we have:

$$\begin{aligned} \dfrac{dz}{dt} &= D\big((S_{in} + X_{in}) - z\big) \Longrightarrow z(t) = (S_{in} + X_{in}) \\ &\quad + \big((s(0) + x(0)) - (S_{in} + X_{in})\big)\mathrm{e}^{-Dt} \end{aligned}$$

It can be derived, as for the minimal model, that the solutions are bounded. The segment:

$$I = \{(s, x) : s \geq 0,\ x \geq 0,\ s + x = S_{in} + X_{in}\}$$

is invariant and attractive.

2.3.1.2. *Equilibrium*

This is the big difference with the minimal model. Here, regardless of D there is at least always a stable equilibrium where *biomass* is strictly positive. There is no "washout". In effect, at equilibrium it follows that:

$$\begin{aligned} D(S_{in} - s) - \mu(s)\,x &= 0 \\[1ex] D(X_{in} - x) + \mu(s)x &= 0 \end{aligned} \qquad [2.23]$$

On the other hand, since at equilibrium we have $S_{in} + X_{in} = s + x$, in the second equation x can be replaced by $S_{in} + X_{in} - s$ which gives:

$$D(X_{in} - ((S_{in} + X_{in}) - s) + \mu(s)(S_{in} + X_{in} - s)$$

that is finally:

$$\mu(s) = D\frac{S_{in} - s}{S_{in} + X_{in} - s} \qquad [2.24]$$

Figure 2.11 shows why, when μ is increasing, the equation:

$$\mu(s) = D\frac{S_{in} - s}{S_{in} + x_{in} - s}$$

has, for any $D > 0$, a unique solution s^* such that $s^* < S_{in}$. It can be shown in the same way that for the minimal chemostat this equilibrium is globally asymptotically stable. In conclusion, "washout" cannot take place when there is a nonzero *biomass* concentration in the input. Which was *a priori* obvious! If the in-flow of liquid contains biomass, there will always be a little bit of it inside the reactor. The minimal chemostat appears as the limit case of this extension when $X_{in} = 0$.

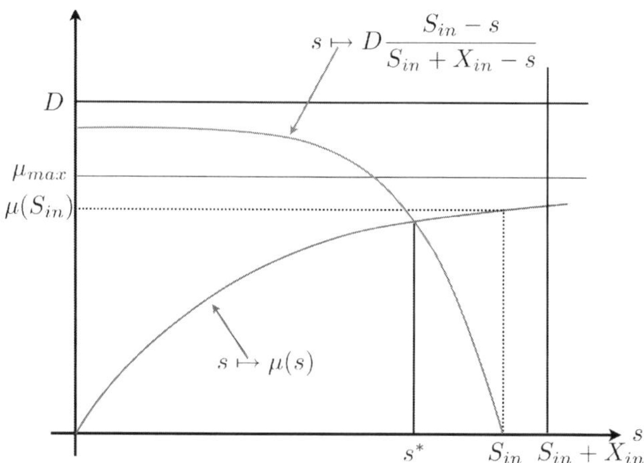

Figure 2.11. *Solution of the equation* $\mu(s) = D\dfrac{S_{in} - s}{S_{in} + X_{in} - s}$. *For a color version of this figure, see www.iste.co.uk/harmand/chemostat.zip*

2.3.2. *Different dilutions*

We have assumed that for a time Δt, the *biomass* that "disappears" from the chemostat is:

$$QV x(t)\Delta t$$

which amounts to an assumption that all the cells that are in the volume that comes out of the reactor when overflowing, and only these, disappear from the *total biomass* in the reactor. Nonetheless, it could be the case that in addition some cells are dying, either a filter retains the cells of a certain size, or still some remain "stuck". In this case, it will be necessary to express the model:

$$\left\{ \begin{array}{rcl} \dfrac{ds}{dt} & = & D(S_{in} - s) - \mu(s)\, x \\[2mm] \dfrac{dx}{dt} & = & (\mu(s) - D_x)\, x \end{array} \right. \qquad [2.25]$$

with:

$$D \neq D_x$$

according to the phenomenon being modeled. The positive orthant and the axis os are always invariant. The big difference with the model [2.2] is that $z = s + x$ no longer satisfies the differential equation $\frac{dz}{dt} = D(S_{in} - z)$. As a result, $\{(s, x) : s + x = S_{in}\}$ is no longer an attractive invariant set. We have however:

$$\frac{dz}{dt} \leq \delta(S_{in} - z)$$

with $\delta = \min\{D, D_x\}$ which is sufficient to show that the solutions are bounded. The "washout" equilibrium $(S_{in}, 0)$ still exists and equilibria with biomass are given by:

$$E_1 = (s^*, x^*)$$

with $\mu(s^*) = D_x$ and $(S_{in} - s^*) > 0$, $x^* = \frac{D}{D_x}(S_{in} - s^*)$. To determine the local stability of a such equilibrium, the Jacobian matrix is calculated:

$$J(E_1) = \begin{bmatrix} -D_s - \mu'(s^*)x^* & -D_x \\[2mm] \mu'(s^*)x^* & 0 \end{bmatrix} \qquad [2.26]$$

The trace of $J(E_1)$ is negative and the determinant positive, therefore the eigenvalues have negative real parts. However, the reader will verify that, contrary to the case of the minimal model in which eigenvalues were always real, they now may eventually be complex; in this case, the return to the equilibrium presents oscillations as it can be seen in Figure 2.12 where we have simulated the model:

$$\begin{cases} \dfrac{ds}{dt} &= 0.1(S_{in} - s) - \dfrac{s}{0.2 + s}\, x \\[2ex] \dfrac{dx}{dt} &= \left(\dfrac{s}{0.2 + s} - 0.7 \right) x \end{cases} \qquad [2.27]$$

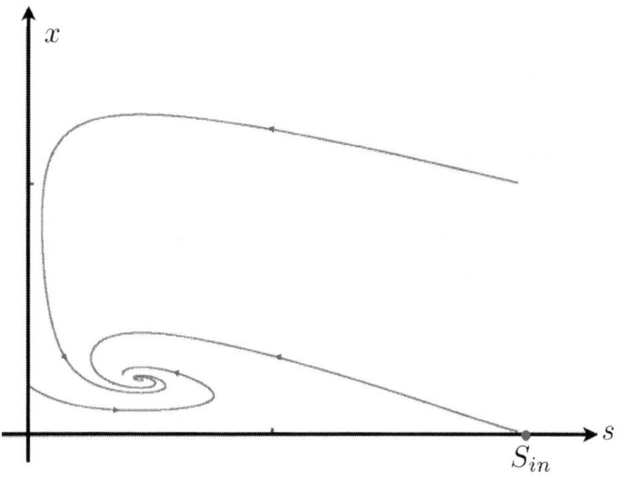

Figure 2.12. *Simulation of [2.27] with $S_{in=2}$. For a color version of this figure, see www.iste.co.uk/harmand/chemostat.zip*

The simulation suggests that the equilibrium is globally stable. In effect, we have the following proposition 2.2:

PROPOSITION 2.2.– If the function μ is of the "Monod type" the equilibrium $E_1 = (s^*, x^*)$ with positive biomass [2.25] is globally asymptotically stable.

Demonstration. We will start with an attempt of demonstration. If we recall the demonstration of proposition 2.1 we must draw the isoclines of [2.25] in order to delimit domains where [2.2] is monotonic. We obtain exactly the same result as in Figure 2.3 since the isoclines are given by the same equations:

– isocline of s: $D(S_{in} - s) - \mu(s)x = 0$;
– isocline of x: $(\mu(s) - D_x)x = 0$;

except that for the model [2.2], we have $D = D_x$, which does not change the figure. We can then conclude, as in the case of the proof of proposition 2.1 that, *a priori* trajectories may spiral around the equilibrium. In proposition 2.1 the convergence toward the equilibrium had been provided by the argument that, in fact, trajectories *cannot spiral* because they are prevented by the presence of the invariant set. In the present case, this argument cannot be used since, as we have just observed, trajectories can actually "spiral". It could then be the case that the presence of a periodic solution surrounding E_1 prevents the stability from being global (see Figure 2.13).

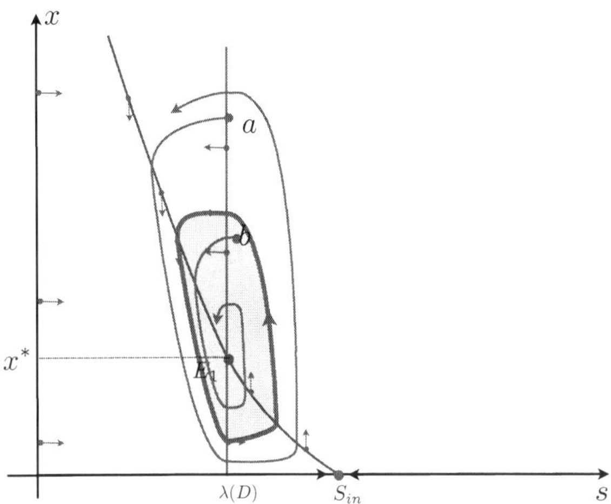

Figure 2.13. *The presence of a cycle could prevent the overall stability. Explanations in section 2.3.2. For a color version of this figure, see www.iste.co.uk/harmand/chemostat.zip*

Thus, if we want to show the global stability, we must find an argument that excludes the existence of a cycle surrounding the equilibrium around which solutions would spiral. To this end, we have to use more subtle arguments than those used so far: the Poincaré-Bendixson theorem (theorem A1.10), the Dulac-Bendixson criterion (theorem A1.11) and the Butler McGehhee theorem (theorem A1.7) which are exposed in section A1.3 in Appendix 1 and that may be omitted during the first reading. proposition 2.2 is a special case of proposition A1.25.

REMARK 2.5.– For the first time, we can observe that the assumption $D = D_x$ greatly simplifies the study. The main reason therefore is the following. We have seen that in the case $D = D_x$, the set:

$$I = \big\{ (s,x) : s \geq 0, \ x \geq 0, \ s + x = S_{in} \big\}$$

is thus an attractive invariant set. Therefore, "In any case, for t large enough, everything appears as if" the system were contained within the set I of dimension one. In fact, the mathematically rigorous justification of "in any case, for t large enough, everything appears as if" is sometimes delicate (see section A1.3.3) where we describe a few traps.

2.3.3. *Density-dependent growth rate and characteristic at equilibrium*

In the minimal model [2.1], the growth rate $\mu(s)$ depends only on the density of the substrate and not on that of the biomass. In this section, we will examine the case (to which we will return in more detail in Chapter 4) where this growth rate is a function $\mu(s, x)$ that also depends on the variable x, which we had already discussed in Chapter 1. In addition, we assume that the dilutions rates of the substrate and of the biomass are different, the chemostat model [2.2] then becomes:

$$
\begin{aligned}
\frac{ds}{dt} &= D(S_{in} - s) - \mu(s, x)x \\
\frac{dx}{dt} &= (\mu(s, x) - D_x)x
\end{aligned}
\tag{2.28}
$$

NOTATION 2.2.– If $f(x_1, \cdots, x_i, \cdots, x_n)$ is a mutivariable function admitting partial derivatives, we denote by $\partial_{x_i} f(x_1, \cdots, x_i, \cdots, x_n)$ the partial derivative with respect to x_i.

We make the following assumptions:

HYPOTHESIS 2.1.–

1) The function μ is defined for s and x positive or zero.

2) We have $\partial_x \mu(s, x) < 0$: for all s, the function $x \mapsto \mu(s, x)$ is strictly decreasing.

3) For any x, the function $s \mapsto \mu(s, x)$ is of the "Monod type", that is zero at 0, bounded and in addition $\partial_s \mu(s, x) > 0$.

Point 2) captures the idea that when the biomass density is very low, we have an intrinsic growth rate (specific to the cell) represented by the function:

$$
s \mapsto \mu_o(s) = \mu(s, 0)
$$

which decreases when the biomass density increases; the reasons may be the production of toxins or the competition for accessing the substrate (or both). Nevertheless, in this model we exclude any form of mutualism that would increase the growth rate: there is inhibition of growth through increase of biomass. We thus study only a particular form of density-dependence called *intra-specific competition*. Point 3) excludes an inhibition caused by an increase in density of the substrate.

Finally, since for all x the function $s \mapsto \mu(s, x)$ is bounded, when s tends to infinity, it tends toward a point that we denote $\mu(\infty, x)$; similarly, we denote $\mu(s, \infty)$ the limit when x tends to infinity.

Possible examples of such functions are:

$$\mu(s, x) = \frac{\mu_{max} s}{k + s} \frac{1}{1 + x^\alpha / l} \qquad [2.29]$$

$$\mu(s, x) = \frac{\mu_{max} s}{\alpha + k x + s} \qquad [2.30]$$

where all parameters are positive numbers.

For [2.29], the species inhibits its own growth by decreasing the maximum of its growth function, since the maximum that is equal to:

$$\sup_{s \geq 0} \mu(s, x) = \frac{\mu_{max}}{1 + x^\alpha / l}$$

decreases when x increases; the semi-saturation constant k remains unchanged. One possible interpretation is as follows. The term $\mu(s)x$ assumes that all individuals of the population have equal access to the substrate but if this is not the case it is $\mu(s)\rho(\cdot)x$ where $\rho(\cdot)$ designates the proportion of individuals that have access to the substrate. When individuals aggregate to form a floc, the substrate struggles to penetrate inside the floc, and thereby only individuals in the floc periphery have access to the substrate. Imagine a (completely unrealistic) situation where biomass x would be made up of a single sphere; in this case the active biomass (with access to the substrate) is a constant thickness layer located on the surface, thus of mass proportional to $x^{2/3}$ and the growth is therefore given by $\mu(s)x^{2/3} = \mu(s)\frac{1}{x^{1/3}}x$. Since this is only valid for significant values of x, it can be corrected as in [2.29].

For [2.30], the species inhibits its own growth by increasing its semi-saturation constant since it is equal to x fixed $\alpha + k x$. In both cases, [2.29] and [2.30], the function μ_o is a Monod function. In the case of [2.30] when $\alpha = 0$, the so-called Contois function is obtained:

$$\mu(s, x) = \frac{\mu_{max} s}{k x + s} \qquad [2.31]$$

where the maximum is equal to μ_{max} and the semi-saturation constant equals $k\,x$ which tends to 0 when x tends to 0 and $\mu_0(s) = \mu_{max}$. Note that the Contois function is not defined nor continuous in $(0,0)$, which raises some mathematical and interpretation problems.

2.3.3.1. *The characteristic at equilibrium*

Unlike the minimal model [2.2] where the isocline of the s is the graph of $s \mapsto \frac{D(S_{in}-s)}{\mu(s)}$ and the isocline of the x is the semi-axis of the s and the half-line $\{(\lambda(D), x) : x \geq 0\}$ in this case the isoclines are defined by the relations:

$$\frac{ds}{dt} = 0 \Leftrightarrow \{(s,x) : s \geq 0; x \geq 0; D(S_{in} - s) - \mu(s,x)x = 0\} \tag{2.32}$$

$$\frac{dx}{dt} = 0 \Leftrightarrow \{(s,x) : s \geq 0; x \geq 0; (\mu(s,x) - D_x)x = 0\} \tag{2.33}$$

where we can no longer immediately solve with respect to s or x due to the presence of the term in x in the function $\mu(s,x)$. In this section, we show how the problem of the determination of equilibria is solved by the introduction of the *characteristic at equilibrium* that will play an important role when we will study the competition of several species in Chapter 4. Consider the second equation of [2.28]:

$$\frac{dx}{dt} = \big(\mu(s,x) - D_x\big)x; \quad x \geq 0 \tag{2.34}$$

where s is considered as a constant parameter. For each value of s, we thus have a particular differential equation whose solutions are being studied.

The following three cases (see Figure 2.14) are possible:

1) $D_x < \mu(s,\infty)$: for any x it follows that $\mu(s,x) > D_x$ therefore [2.34] has the unique equilibrium 0, it is unstable and solutions tend to $+\infty$;

2) $\mu(s,\infty) < D_x < \mu(s,0)$: there exists a unique x denoted $\psi(s)$ such that $\mu(s,\psi(s)) = D_x$; then $\psi(s)$ is a globally asymptotically stable equilibrium;

3) $\mu(s,0) < D_x$: for any x, it follows that $D_x > \mu(s,x)$ therefore [2.34] has the unique equilibrium 0, it is globally asymptotically stable.

It follows that for a fixed s, the system [2.34] has a unique globally asymptotically stable equilibrium (which is $+\infty$ in case 1).

DEFINITION 2.2.– *The application that associates the unique stable equilibrium of [2.34] with s, which we denote $s \mapsto \psi(s)$, is called* characteristic at equilibrium *of the system [2.28].*

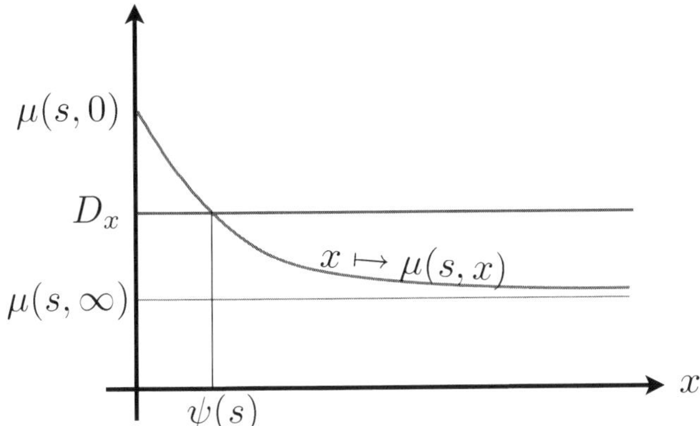

Figure 2.14. *Determination of the equilibria of [2.34]. For a color version of this figure, see www.iste.co.uk/harmand/chemostat.zip*

We continue to define the *break-even concentration* $s = \lambda(D_x)$ by $\mu(\lambda(D_x),0) = D_x$; we can see that the characteristic at equilibrium is a function which is zero for $0 \leq s \leq \lambda(D_x)$ and then increasing. Without additional assumptions, we cannot decide if it is defined for all s and bounded, or if it is defined for all s and tends to infinity, or finally if it tends to infinity for a finite value of s; all cases are possible.

The reader will verify that when $\mu(s,x)$ is defined by the function [2.29], for $D_x < \mu_{max}$ the function $s \mapsto \psi(s)$ is defined by:

$$
\begin{aligned}
s \leq \lambda(D_x) &\implies \psi(s) = 0 \\
s \geq \lambda(D_x) &\implies \psi(s) = \left(l\left(\frac{1}{D_r} \frac{\mu_{max}\, s}{k+s} - 1 \right) \right)^{\frac{1}{\alpha}}
\end{aligned}
\qquad [2.35]
$$

with:

$$
\lambda(D_x) = \frac{kD_x}{\mu_{max} - D_x}
$$

It is a function defined for all s and bounded (Figure 2.15, left). We see that when l tends to infinity, that is when $\mu(s,x)$ depends less and less on x, the graph of ψ "straightens" and tends toward the vertical segment $\mu(s,0) = D_x$.

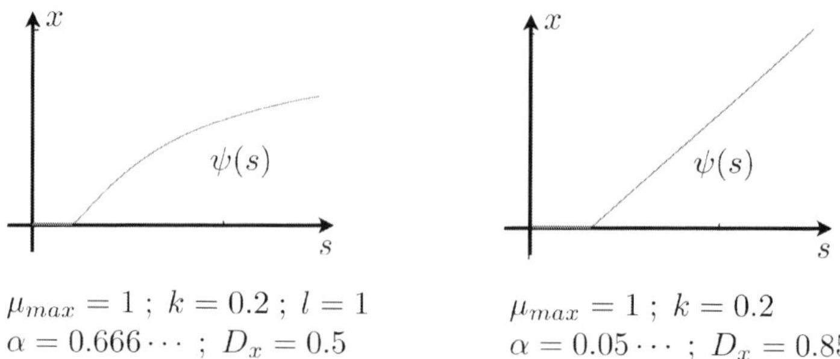

$$\mu_{max} = 1 \; ; \; k = 0.2 \; ; \; l = 1$$
$$\alpha = 0.666 \cdots \; ; \; D_x = 0.5$$

$$\mu_{max} = 1 \; ; \; k = 0.2$$
$$\alpha = 0.05 \cdots \; ; \; D_x = 0.85$$

Figure 2.15. *Graph of [2.35] (left); graph of [2.36] (right). For a color version of this figure, see www.iste.co.uk/harmand/chemostat.zip*

We will also verify that when $\mu(s, x)$ is defined by [2.30], the function $\psi(s)$ is defined by:

$$
\begin{aligned}
s \leq \lambda(D_x) &\implies \psi(s) = 0 \\
s \geq \lambda(D_x) &\implies \psi(s) = \frac{(\mu_{max} - D_x)s - \alpha}{k\,D_x}
\end{aligned}
\qquad [2.36]
$$

with:

$$\lambda(D_x) = \frac{\alpha}{\mu_{max} - D_x}$$

For $s \geq \lambda(D_x)$, it is a linear function whose graph is a straight line (Figure 2.15, right).

2.3.3.2. *Equilibria*

Let us go back to the system [2.28]. We prove exactly as in section 2.3.2 that the solutions are bounded and that the horizontal axis is invariant. For equilibria, we have the following proposition 2.3:

PROPOSITION 2.3.– Denote by (s^*, x^*) the coordinates of the unique point of intersection of the graph of $s \mapsto \psi(s)$ with the line segment: $\{(s, x) : D(S_{in} - s) - D_x x = 0 \; ; \; s \geq 0 \; ; \; x \geq 0\}$.

1) the point of coordinates (s^*, x^*) is an equilibrium of [2.28];

2) if $S_{in} \leq \lambda(D_x)$ the point: $E_o = (s^*, x^*) = (S_{in}, 0)$ is a locally exponentially stable equilibrium. This is the only (washout) equilibrium of [2.28];

3) If $S_{in} > \lambda(D_x)$:

   - the point: $E_o = (S_{in}, 0)$ is an unstable (washout) equilibrium (a saddle);

   - the point: $E_1 = (s^*, x^*)$ is an equilibrium (with positive biomass) locally exponentially stable.

Demonstration. Point 1) is obvious.

– if the break-even concentration $\lambda(D_x)$ is smaller than S_{in}, the Jacobian matrix in (s^*, x^*):

$$J(E_1) = \begin{bmatrix} -D_s - \partial_s\mu(s^*, x^*)x^* & -D_x \\ \\ \partial_s\mu(s^*, x^*)x^* & \partial_x\mu(s^*, x^*)x^* \end{bmatrix} \qquad [2.37]$$

has a negative trace and a positive determinant due to assumptions 2.1. The Jacobian matrix in $(S_{in}, 0)$ is:

$$J(E_0) = \begin{bmatrix} -D_s & -\mu(S_{in}, 0) \\ \\ 0 & \mu(S_{in}, 0) - D_x \end{bmatrix} \qquad [2.38]$$

defines a saddle because $\lambda(D_x) < S_{in} \implies D_x < \mu(S_{in}, 0)$;

– if the break-even concentration $\lambda(D_x)$ is greater than S_{in}, the only equilibrium is the washout equilibrium and it is exponentially stable because now $\mu(S_{in}, 0) - D_x < 0$.

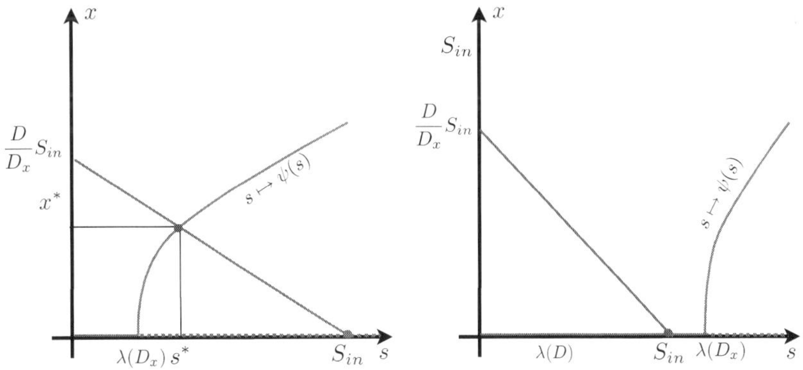

Figure 2.16. *The equilibria of [2.28]. For a color version of this figure, see www.iste.co.uk/harmand/chemostat.zip*

In fact, the equilibria (s^*, x^*), when $\lambda(D_x) \leq S_{in}$ and $(S_{in}, 0)$ when $\lambda(D_x) > S_{in}$ are globally stable. We have in effect the following theorem:

THEOREM 2.1.– The equilibrium (s^*, x^*) defined in proposition 2.3 is globally asymptotically stable.

Demonstration. See A1.3.4.

2.3.4. *Yield depending on the density of the substrate*

The "minimal model" and the three extensions that we have proposed have the peculiarity that they never show any limit cycle: solutions always tend toward an equilibrium. However, in these three extensions the yield Y was constant. If we no longer make this assumption, limit cycles may appear. Since the study of these models is more delicate, we merely present simulations in this section. More information on this topic can be found in the appendices, in particular in exercise A1.2 and in section A1.3.

We are going to observe what happens in the model proposed in the article [ARI 03]:

$$\begin{aligned}
\frac{ds}{dt} &= (1 - s) - \frac{2s}{0.58 + s}\frac{10}{1 + 46s^2}\,x \\
\frac{dx}{dt} &= \left(\frac{2s}{0.58 + s} - D_x\right)x
\end{aligned}$$

[2.39]

The yield is the function:

$$y(s) = \frac{1 + 46s^2}{10}$$

It is an increasing function of the substrate concentration which can have the following interpretation: the more abundant the substrate is, the less bacteria need to "move around" to find some of it and, consequently the part of substrate dedicated to growth is relatively larger. The fact that the yield is greater than 1 is no reason to worry us insofar as units are not specified; on the other hand, we do not intend here any realism and we just want to describe a mathematical phenomenon. The isocline $\frac{ds}{dt} = 0$ is the graph of the function:

$$\pi(s) = \frac{(1 - s)(0.58s)(1 + 46s^2)}{20\,s}$$

Since the Monod function $s \mapsto \frac{2s}{0.58+s}$ is strictly monotone, for $D_x < 2$ there exists a unique s^* such that $\left(\frac{2s^*}{0.58+s^*} - D_x\right) = 0$ and for $D_x < 1.266\cdots$ this s^* is smaller than $S_{in} = 1$. Therefore, for $D_x < 1.266\cdots$ there exists a unique equilibrium with positive biomass: $E_1 = (s^*, x^*)$ where $x^* = \pi(s^*)$. One will verify that this equilibrium is locally exponentially stable if E_1 is located inside a decreasing section of the graph of π, unstable otherwise (see Figure 2.17).

When s increases, the graph of $s \mapsto \pi(s)$ decreases from ∞, passes through a minimum, increases up to a maximum and then decreases again to cancel out for $s = S_{in}$ (see Figure 2.17). The maximum is reached for $s_M = 0.579514\cdots$. The value of D_x for which the $s^* = s_M$ is $D_x^0 = 0.999580\cdots$. Thus:

– for $D_x > 0.999580\cdots$, the equilibrium E_1 is in a decreasing section of the graph of π: it is locally exponentially stable;

– for $D_x < 0.999580\cdots$, the equilibrium E_1 is inside an increasing section of the graph of π: it is unstable.

We are going to observe simulations of the phase portrait for values close to D_x^0.

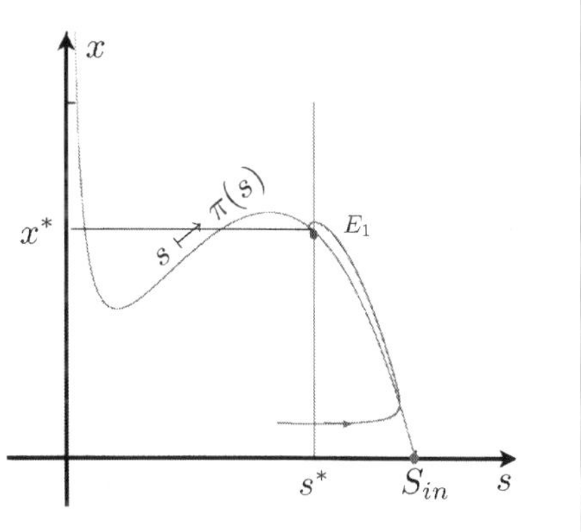

<table>
<tr><td>

The figure on the right shows in black the isocline $s \mapsto \pi(s)$ of the model [2.39] as well as the straight line $\{(s,x) : s = s^*; x > 0\}$. The intersection of these two isoclines defines the equilibrium E_1; this equilibrium, being located inside a section with negative slope of the graph of π, is locally exponentially stable. In red, we have simulated a trajectory; it tends to E_1.

</td></tr>
</table>

Figure 2.17. *Model [2.39]. For a color version of this figure, see www.iste.co.uk/harmand/chemostat.zip*

COMMENTS ON FIGURE 2.18.– Four simulations have been represented for increasing values of D_x:

– a) we have $D_x = 0.9$; the equilibrium E_1 (the red dot) is unstable. The trajectories originate from a point close to the unstable equilibrium "spiral" and move away from E_1 to end up accumulating on a periodic solution (in blue); the trajectories originating from an external point wind up around the periodic solution that is thus a *stable limit cycle.* Note that we have not demonstrated that this cycle is unique as the simulation seems to show;

– b) we have $D_x = 0.99$, the equilibrium E_1 (the red dot) is unstable. The limit cycle has become therefrom a bit distorted;

– c) we have $D_x = 1.01$, the equilibrium E_1 (the blue dot) is stable. A solution has been represented with an initial condition far from the equilibrium. The corresponding solution spirals slowly toward equilibrium;

– d), we have $D_x = 1.1$, the equilibrium E_1 (the blue dot) is stable. Solutions tend much faster toward equilibrium.

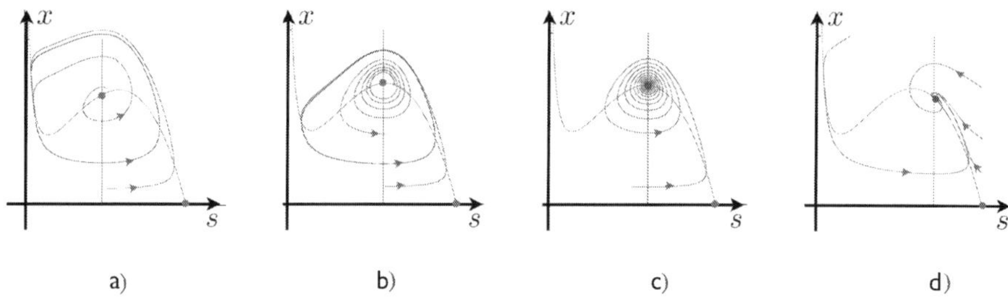

Figure 2.18. *Model [2.39]:*
$D_x = 0.900\,(a);\ 0.990\,(b);\ 1.010\,(c);\ 1.100\,(d).$ *For a color version of this figure, see www.iste.co.uk/harmand/chemostat.zip*

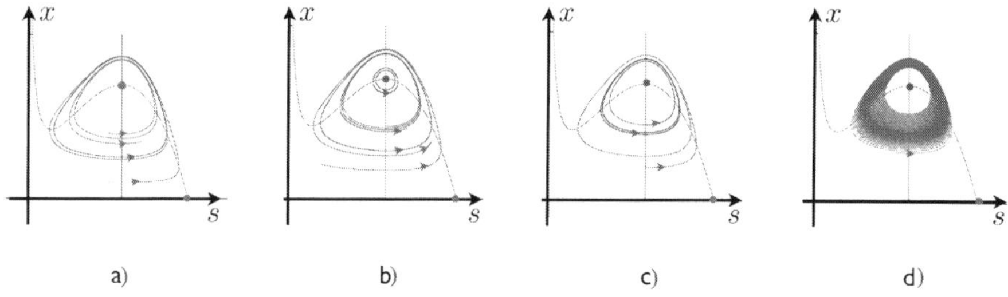

Figure 2.19. *Model [2.39]:*
$D_x = 0.999\,(a);\ 1.000\,(b);\ 1.0009\,(c);\ 1.001\,(d).$ *For a color version of this figure, see www.iste.co.uk/harmand/chemostat.zip*

COMMENTS ON FIGURE 2.19.– As previously, we observe four simulations for increasing values of D_x, this time very close of the value 0.999580 where the stability of the equilibrium is reversed. The variation of D_x is very small (2×10^{-3}):

– a) we have $D_x = 0.999$; the equilibrium E_1 (the red dot) is always unstable but instability is very weak; close to the equilibrium the trajectories appear as periodic;

– b) we have $D_x = 1.000$; the equilibrium E_1 (blue dot) is now stable but the (large) periodic solution has also remained locally stable, which has been possible due to the birth of a small unstable periodic solution surrounding E_1. The equilibrium E_1 is locally stable and its basin of attraction is the interior of the (small) periodic solution; the basin of attraction of the large periodic solution is the exterior of the small periodic solution;

– c) we have $D_x = 1.0009$. When D_x is increased, it is observed that the small periodic solution increases while the large one decreases and for the value 1.0009, they are almost similar;

– d) we have $D_x = 1.001$. The equilibrium E_1 is stable. The solution represented spirals very slowly toward equilibrium; a strengthening of the red color can be observed which corresponds to more closely joined spiral loops, therefore to a lower velocity of convergence to the equilibrium. There is thus a slowdown in the velocity of convergence to the equilibrium and then a re-acceleration.

2.4. Bibliographic notes

The first comprehensive mathematical treatment of the equations of the chemostat is due to Spicer [SPI 55]. The book by Smith and Waltman [SMI 95] published in 1995 contains most of the mathematical developments published at this time. In addition to the treatment of the minimal model, it contains developments that are not addressed in this book, in particular: spatially structured models (gradostat), periodic dilution models, models with internal storage (that decouple the activity of harvesting the substrate from that of the creation of biomass), and size-structured models. This book, available on the Internet, is a great classic.

In 1989, Arditi and Ginzburg [ARD 89] contributed with a change in perspective decisive in the manner in which the relationship "resource-consumer" (or prey-predator) is modeled in the interactions between populations. Instead of the traditional term:

$$\mu(s)x$$

where the capture rate of the resource $\mu(\cdot)$ is a function only of the substrate concentration, they have advocated the use of the *ratio-dependent* model:

$$\mu(s/x)x$$

where, this time, the capture rate of the resource is a function of the ratio s/x, that is of the amount of resource available per individual and not the quantity in absolute terms. This modification that quite substantially changes the qualitative properties of models has been heatedly discussed before being accepted. We do not account for these discussions that relate more to theoretical ecology and take us away from the subject of this book. The reader who wishes to address works on the chemostat considering the general framework of population dynamics may consult the recent book by Arditi and Ginzburg [ARD 12]. Section 2.3.3 of this book, where we introduce density-dependent growth rates, is intended to take this new vision into account, the ratio-dependency being a particular case of density dependence.

For the model with variable yield, the reader may consult the bibliographic notes of Appendix 1, section A1.5.

3

Competitive Exclusion

In this chapter, we consider several *species* of microorganisms in a chemostat, which are in *competition* for a limiting resource for their growth. We write a generalization to several species of the chemostat model presented and studied in the previous chapter:

$$
\begin{cases}
\dfrac{ds}{dt} = D(S_{in} - s) - \displaystyle\sum_{i=1}^{n} \mu_i(s)x_i \\[4mm]
\dfrac{dx_i}{dt} = \mu_i(s)x_i - Dx_i \quad i = 1 \cdots n
\end{cases}
\tag{3.1}
$$

This formalization is the simplest extension and the more natural of the monospecific chemostat model because it considers that the dynamics of each strain i, of concentration x_i, is governed as if it were alone in an environment where the resource is in concentration s, without any *direct* interaction with the other species. The interactions between species are only *indirect*, sharing the same limiting resource. The dynamics of the resource is therefore dependent on the sum of the contributions of each species i during consumption of the shared substrate. As in the previous chapter, we assume, without loss of generality, that yield coefficients Y_i are equal to 1 (remember that by choosing the measure unit of biomass x_i, it is possible to replace x_i/Y_i by x_i).

We first begin by determining the possible steady-states of the system [3.1] in the positive orthant.

3.1. The case of monotonic growth functions

We assume here that all the functions μ_i are strictly increasing.

3.1.1. *Steady states*

We recall the definition of break-even concentration introduced in the previous chapter.

DEFINITION 3.1.– *For a fixed dilution rate D, $\lambda_i(D)$ designates break-even concentration for species i, the concentration s (if it exists) that verifies $\mu_i(s) = D$. If there is no solution to this equation, the convention is to define $\lambda_i(D) = +\infty$.*

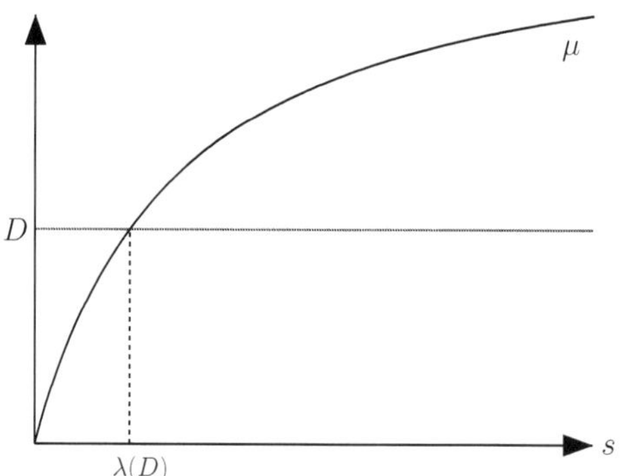

Figure 3.1. *Break-even concentration*

For a Monod function:

$$\mu(s) = \frac{\mu_{\max}\, s}{K_S + s}$$

we recall the expression of the break-even concentration for $D < \mu_{\max}$:

$$\lambda(D) = \frac{D\,K_S}{\mu_{\max} - D} \qquad [3.2]$$

and $\lambda(D) = +\infty$ if $D \geq \mu_{\max}$. It should be denoted that $s_i^\star = \lambda_i(D)$ corresponds to the value of s at non-washout steady-state of species i in the mono-specific chemostat model studied in the previous chapter, as soon as we have $s_i^\star < S_{in}$.

3.1.2. *Possible steady-states*

We are studying the possibilities of simultaneously canceling out all the members of the right-hand side of the system [3.1] according to the number of species existing at steady-state:

– there are no species present at steady-state ($x_j^\star = 0$ for all j). Then, it necessarily follows that $s^\star = S_{in}$. This corresponds to the *washout* steady-state, which we denote by E_0;

– only one species is present at steady-state: all the $x_j^\star$ are equal to zero except for a certain $j = i$, which is a steady-state that we denote E_i. We are left with the monospecific model of the chemostat studied in the previous chapter: $\lambda_i(D) < S_{in}$ is the condition for the existence of such a (positive) steady-state. The values at steady-state of variables s and x_i are then $s_i^\star = \lambda_i(D)$ and $x_i^\star = S_{in} - s_i^\star$;

– at least two species are present at steady-state, admittedly for indices i and j. It corresponds to a *coexistence* steady-state. This requires that $\dot{x}_i = 0$ and $\dot{x}_j = 0$ for $x_i > 0$ and $x_j > 0$, which implies the existence of a value $s^\star$ that verifies both $\mu_i(s^\star) = D$ and $\mu_j(s^\star) = D$, which is tantamount to writing $s^\star = \lambda_i(D) = \lambda_j(D)$.

3.1.2.1. *The non-generic character of coexistence steady-states*

Competitive exclusion at steady-state, that we will expose in more detail a bit further on, is based on the fact that the condition $\lambda_i(D) = \lambda_j(D)$ required for the existence of coexistence steady-state is very constraining and never satisfied in practice. In fact, this condition means first of all that the graphs of the functions μ_i and μ_j cross each other outside 0, precisely at $s^\star$. This property can be verified for example with Monod expressions (we leave as an exercise for the reader to show that the graphs of the two distinct Monod functions can actually intersect each other, and that moreover, there is at least an intersection outside 0). However, this condition also implies that the dilution rate D exactly corresponds to the common value $\mu_i(s^\star) = \mu_j(s^\star)$. If it is considered that the intersections between the graphs of growth functions can only occur for isolated values of s, then the corresponding values of the dilution rate D which would yield $\mu_i(s) = \mu_j(s) = D$ are also isolated values. In practice, this means that it should be necessary to maintain the dilution rate D very accurately at one of these isolated values so that a coexistence steady-state is created; furthermore, the smallest change in the dilution rate would make all the possible coexistence steady-states disappear. Such steady-states are almost impossible to obtain in practice. Thus, apart from these "exceptional" values of D, it can be inferred that the model [3.1] does not allow steady-states with more than one species. Similarly, we consider that the strict equality $\lambda_i(D) = S_{in}$, which corresponds to the particular case where the steady-state E_i coincides with E_0, is an all the more exceptional situation which is never encountered in practice.

From now on, we will only consider "non-exceptional situations", namely for which all the $\lambda_i(D)$ are distinct and different from S_{in}.

3.1.3. *Local stability of washout steady-state*

We study the stability of washout steady-state E_0 by making use of the system Jacobian matrix, which is written in E_0:

$$J(E_0) = \begin{bmatrix} -D & -\mu_1(S_{in}) & \cdots & -\mu_n(S_{in}) \\ \hline 0 & \mu_1(S_{in}) - D & & (0,0)0 \\ \vdots & & \ddots & \\ 0 & (0,0)0 & & \mu_n(S_{in}) - D \end{bmatrix}$$

Since the matrix is triangular the eigenvalues are the diagonal values. For example, when $\mu_i(S_{in}) < D$ for all $i = 1 \cdots n$, the steady-state E_0 is locally exponentially stable. The functions μ_i being strictly increasing, this condition is equivalent to require that $\lambda_i(D) > S_{in}$ for all i. E_0 is then the only steady-state of the system. On the contrary, as soon as there is an index j such that $\lambda_j(D) < S_{in}$, the steady-state E_j exists and it necessarily follows that $\mu_j(S_{in}) > D$. The steady-state E_0 is then unstable.

3.1.3.1. *Local stability of steady-state outside washout*

First of all, consider the steady-state E_1, under the condition of existence $\lambda_1(D) < S_{in}$, and let us examine the eigenvalues of the Jacobian matrix:

$$J(E_1) = \begin{bmatrix} -D - \mu_1'(s_1^\star)x_1^\star & -D & -\mu_n(s_1^\star) & \cdots & -\mu_n(s_1^\star) \\ \mu_1'(s_1^\star)x_1^\star & 0 & 0 & \cdots & 0 \\ \hline & & \mu_2(s_1^\star) - D & & 0 \\ & 0 & & \ddots & \\ & & 0 & & \mu_n(s_1^\star) - D \end{bmatrix}$$

This matrix can be decomposed in a block-wise fashion, where the 2×2 submatrix made of the first two rows and the first two columns corresponds to that of the monospecific chemostat model for the non-washout steady-state of species 1, whose two eigenvalues are strictly negative according to the study of the previous chapter. The other eigenvalues are $\mu_2(s_1^\star) - D$, ..., $\mu_n(s_1^\star) - D$. Since the function μ_i is strictly increasing, the requirement that the eigenvalue $\mu_j(s_1^\star) - D$ has to be strictly negative is equivalent to requiring that $\lambda_1(D) < \lambda_j(D)$. Thus steady-state E_1 is locally exponentially stable as soon as we obtain $\lambda_1(D) < \lambda_j(D)$ for any $j \neq 1$, and unstable if there exists $j \neq 1$ such that $\lambda_j(D) < \lambda_j(D)$.

More generally, by swapping indices 1 and i, a steady-state E_i, under the condition of existence $\lambda_i(D) < S_{in}$, is locally exponentially stable when $\lambda_i(D) < \lambda_j(D)$ for all $j \neq i$, and unstable if there exists $j \neq i$ such that $\lambda_j(D) < \lambda_i(D)$.

EXERCISE 3.1.– Express the system [3.1] in coordinates $(m, x_1, \cdots, x_n)$ instead of $(s, x_1, \cdots, x_n)$ where $m = s + \sum_{i=1}^{n} x_i$. In these coordinates, express the Jacobian matrices for steady-states E_0 and E_1, then redo the calculation of eigenvalues to determine the stability of steady-states.

3.2. Competitive exclusion at steady-state

Consider n species with distinct break-even concentration, and we assume that:

$$\lambda_1(D) < \lambda_2(D) < \cdots < \lambda_n(D)$$

(we exclude exceptional situations for which there would exist indices $i \neq j$ such that $\lambda_i(D) = \lambda_j(D)$). The conditions for the existence and local stability of different steady-states can be summarized by the following table.

	$S_{in} < \lambda_1(D)$	$\lambda_1(D) < S_{in} < \lambda_2(D)$	$\lambda_2(D) < S_{in} < \lambda_3(D)$	$\cdots$	$\lambda_n(D) < S_{in}$
E_0	stable	unstable	unstable	$\cdots$	unstable
E_1	does not exist	stable	stable	$\cdots$	stable
E_2	does not exist	does not exist	unstable	$\cdots$	unstable
$\vdots$					
E_n	does not exist	does not exist	does not exist	$\cdots$	unstable

Table 3.1. *Summary of the various possible situations according to the respective position of the parameter S_{in} with respect to break-even concentration $\lambda_i(D)$*

This allows us to state a mathematical result about competitive exclusion *at steady-state.*

3.2.1. *Statement*

PROPOSITION 3.1.– Assume that the functions μ_i ($i = 1 \cdots n$) are strictly increasing and verify $\mu_i(0) = 0$. When break-even concentrations $\lambda_i(D)$ are distinct and different from S_{in}, the following properties are satisfied:

– if all break-even concentrations are greater than S_{in}, the washout E_0 is the only steady-state of the system [3.1], which is then locally exponentially stable;

– for each break-even concentration $\lambda_i(D)$ smaller than S_{in}, there exists a steady-state E_i. Only the steady-state $E_{i^\star}$ corresponding to the smallest break-even concentration is locally exponentially stable, and the other steady-states E_i (when they exist) are unstable. There is no coexistence steady-state;

It should be noted that this statement addresses only the local stability of steady-states. The issue of global stability will be covered in section 3.3.

EXERCISE 3.2.– Propose and demonstrate an extension of proposition 3.1, when death rates are different:

$$
\begin{cases}
\dfrac{ds}{dt} = D(S_{in} - s) - \displaystyle\sum_{i=1}^{n} \mu_i(s)x_i \\[2em]
\dfrac{dx_i}{dt} = \mu_i(s)x_i - D_i x_i \quad i = 1 \cdots n
\end{cases}
\qquad [3.3]
$$

Indication: consider the break-even concentrations of $\lambda_i(D_i)$ instead of $\lambda_i(D)$ for each species i. As previously, we consider that having equalities $\lambda_i(D_i) = S_{in}$ or $\lambda_i(D_i) = \lambda_j(D_j)$ for j different from i are exceptional situations.

3.2.2. *Species at steady-state according to the dilution rate*

Notice that when the graphs of the functions μ_i cross each other, the index $i^\star$ that achieves the smallest break-even concentration can change by modifying the value of D. We illustrate this property with an example with two species whose growth kinetics have the following Monod expressions:

$$
\mu_1(s) = \frac{(2/5)s}{1/2 + s}, \quad \mu_2(s) = \frac{(1/2)s}{1 + s}
$$

By using the expression [3.2], it follows that $D = 1/5$:

$$
\lambda_1(1/5) = \frac{1}{2} < \lambda_2(1/5) = \frac{2}{3}
$$

whereas for $D = 7/20$, we have:

$$\lambda_1(7/20) = \frac{7}{2} > \lambda_2(7/20) = \frac{7}{3}$$

as illustrated in Figure 3.2 on the interval $[0, 5]$. As in the previous chapter, we consider the *operating diagram* that summarizes the different possibilities of stable steady-state(s) based on the values of the operating parameters D and S_{in}. Therefore, in Figure 3.3, three regions are defined:

– in pink: E_0 is the only steady-state of the system since we have: $\mu_1(S_{in}) < D$ and $\mu_2(S_{in}) < D$. In addition, we have seen that E_0 is a stable equilibrium;

– in blue: steady-state E_1 exists since we have $\lambda_1(D) < S_{in}$. When E_2 exists, it follows that $\lambda_1(D) < \lambda_2(D)$. According to proposition 3.1, E_1 is the only stable steady-state;

– in gray: in this region, the steady-state E_2 exists and when E_1 exists, it follows that $\lambda_2(D) < \lambda_1(D)$. According to proposition 3.1, E_2 is the only stable steady-state.

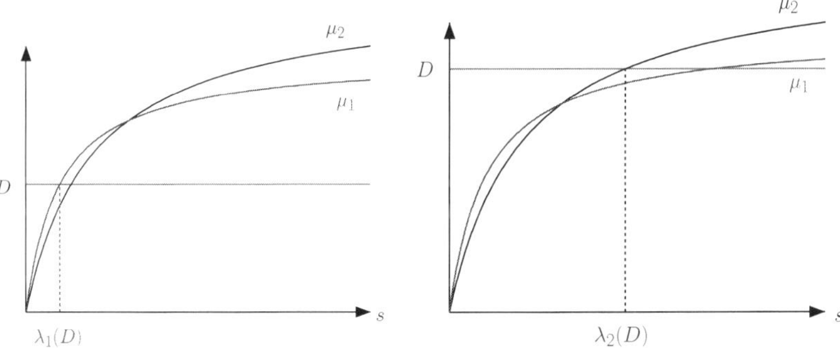

Figure 3.2. *Example in which depending on the value of D, it is a different species that has the smallest break-even concentration ($D = 1/5$ on the left, $D = 7/20$ on the right). For a color version of this figure, see www.iste.co.uk/harmand/chemostat.zip*

3.2.3. *Dynamics of proportions between species*

In the competition between species, proposition 3.1 characterizes the species that can survive in a stable manner at steady-state, but does not give any information about the velocity at which this "winning" species excludes the other species. To simplify, we consider the model [3.1] with two species only, and assume:

$$\lambda_1(D) < \lambda_2(D) \tag{3.4}$$

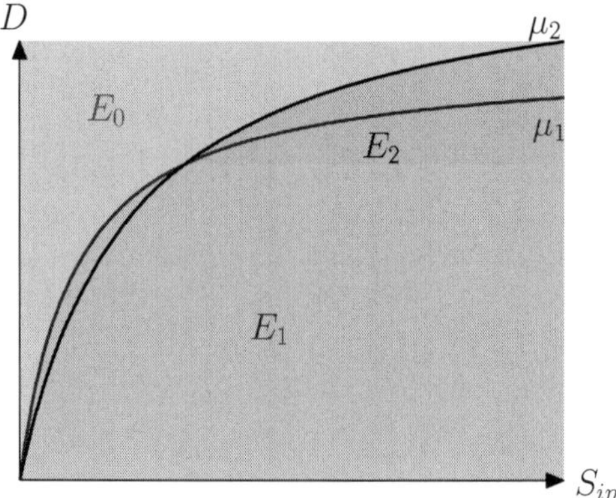

Figure 3.3. *Operating diagram of the stability of steady-states (in pink: E_0 is the only stable steady-state, in blue: E_1 is the only stable steady-state, in gray: E_2 is the only stable steady-state). For a color version of this figure, see www.iste.co.uk/ harmand/chemostat.zip*

To analyze the behavior of the solutions of the system, it is convenient to proceed to the following change of variables:

$$(s, x_1, x_2) \longrightarrow X = (s, b, p)$$

where b and p designate, respectively, total biomass and the proportion of the first species:

$$b = x_1 + x_2, \quad p = \frac{x_1}{b}$$

A simple calculation, which we encourage the reader to verify, shows that in these new variables the system is:

$$\begin{cases} \dfrac{ds}{dt} = & D(S_{in} - s) - (p\mu_1(s) + (1-p)\mu_2(s))\, b \\[2mm] \dfrac{db}{dt} = & (p\mu_1(s) + (1-p)\mu_2(s))\, b - Db \\[2mm] \dfrac{dp}{dt} = & p(1-p)(\mu_1(s) - \mu_2(s)) \end{cases} \qquad [3.5]$$

and the steady-state E_1 is written using these variables $X = (s_1^\star, x_1^\star, 1)$.

3.2.3.1. *The case of "close" species*

When the two species are functionally "close" to each other, in other words, when their growth functions verify:

$$\mu_1(s) = \mu_2(s) + \varepsilon(s)$$

where ε is a function that takes "small" values, the system [3.5] is written as:

$$\begin{cases} \dfrac{ds}{dt} &= D(S_{in} - s) - (\mu_1(s) - (1-p)\varepsilon(s))\, b \\[2mm] \dfrac{db}{dt} &= \mu_1(s)b - Db \\[2mm] \dfrac{dp}{dt} &= p(1-p)\varepsilon(s) \end{cases}$$

Since the values of the function ε are assumed to be small, the dynamics of variable p is "slow", compared to the dynamics of the variables s and b. Thus, the proportion of the first species (which is the winner of the competition under assumption 3.4) will evolve slowly when the term $\varepsilon(s)$ is small, which is reflected by closely similar growth curves of the two species. On the other hand, the subsystem in (s, b) is almost decoupled from the variable p and can be approximated by means of dynamics with one species:

$$\begin{cases} \dfrac{ds}{dt} &= D(S_{in} - s) - \mu_1(s)b \\[2mm] \dfrac{db}{dt} &= \mu_1(s)b - Db \end{cases}$$

Finally, the dynamics [3.5] present two time scales: this corresponds to a so-called "slow-fast" system.

3.2.3.2. *Behavior in the neighborhood of the steady-state*

Now consider more general situations where the growth functions μ_i are not necessarily similar for all s. When the variables of the system [3.5] are in the neighborhood of the steady-state E_1, a development limited to the first order of the dynamics of the variable p is written as:

$$\begin{aligned} \dot{p} &= p(1-p)(\mu_1(s) - \mu_2(s)) \\ &= \left((1-p) - (1-p)^2\right)(\mu_1(s) - \mu_2(s)) \\ &\simeq (1-p)(\mu_1(s_1^\star) - \mu_2(s_1^\star)) \end{aligned}$$

For a small perturbation around E_1, that is for X close to $X^\star$ at time 0, it follows that:

$$p(t) \simeq 1 + (p(0) - 1)e^{-\nu t}, \quad t > 0$$

where ν is defined by:

$$\nu = \mu_1(s_1^\star) - \mu_2(s_1^\star) = D - \mu_2(\lambda_1(D))$$

(positive number under the assumption [3.4]). We thus see that the proportion p returns to 1 all the more slowly since the number ν is small. It should be observed that having a small ν does not imply that the functions μ_i be closely similar for all s as previously seen, but only close about the value $\lambda_1(D)$. We illustrate this behavior with a simulation:

– by establishing the characteristics of the first species, the dilution rate and an initial condition close to the steady-state E_1;

– by considering several possible growth curves for the second species, more or less close to the first, while respecting condition [3.4].

EXAMPLE 3.1.– We choose the Monod expressions:

$$\mu_1(s) = \frac{(2/5)s}{1/2 + s} \quad \text{and} \quad \mu_2(s) = \frac{(1/2)s}{3 + s} \quad \text{or} \quad \frac{(1/2)s}{6/5 + s}$$

By using expression [3.2], it yields for $D = 3/10$:

$$\lambda_1(3/10) = \frac{3}{2} < \lambda_2(3/10) = \frac{9}{2} \quad \text{or} \quad \frac{9}{5}$$

and we determine:

$$\nu = D - \mu_2(\lambda_1) = \frac{2}{15} \quad \text{or} \quad \frac{1}{45}$$

Figure 3.4 shows the graph of the functions μ_i and simulations for $S_{in} = 5$ as well as the initial condition $(s, b, p)(0) = (1.4, 3.4, 0.9)$ close to steady-state $(\lambda_1(D), S_{in} - \lambda_1(D), 1) = (1.5, 3.5, 1)$. In the second case, the graphs of the functions μ_i intersect one another, which gives the possibility to have a small number ν for values of the dilution rate D close to the common value taken by these functions in the intersection. It is thus observed that when ν is small, the system takes longer to separate from the second species, while the total biomass and the substrate have similar time evolutions.

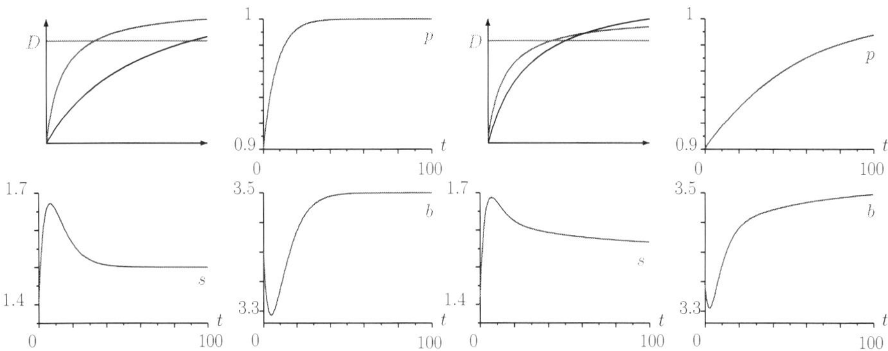

Figure 3.4. *Simulations where $\nu = 2/15 \simeq 0.13$ (on the left); $\nu = 1/45 \simeq 0.02$ (on the right). For a color version of this figure, see www.iste.co.uk/harmand/chemostat.zip*

3.2.4. *Conclusion*

In the previous sections, we have stressed out several times the exceptional nature of obtaining for two distinct species i and j the equality $\lambda_i(D) = \lambda_j(D)$. Nonetheless, it may occur that we could be "close" to such a situation, in the sense that the numbers $\lambda_i(D)$ and $\lambda_j(D)$ are close to each other (without being rigorously equal):

– when the functions μ_i, μ_j are "close", and this is for any value of s;

– when the graphs of the functions μ_i, μ_j intersect one another for a value close to the dilution rate D.

3.3. Global stability

As we have already stressed, proposition 3.1 gives only a condition for the local stability of the steady-state $E_i^\star$. At the price of changing indices, we assume that $\lambda_1(D)$ is the smallest break-even concentration (and is less than S_{in} to ensure the existence of the steady-state E_1 in the positive quadrant). We are going to show the following global result.

PROPOSITION 3.2.– Assume the condition:

$$\lambda_1(D) < \min\left(\min_{i>1} \lambda_i(D), S_{in}\right)$$

is verified. Then for any initial condition $(s(0), x_1(0), \cdots, x_n(0))$ such that $x_i(0) > 0$ for $i = 1 \cdots n$, the solution of the system [3.1] verifies:

$$\lim_{t \to +\infty} (s(t), x_1(t), x_2(t), \cdots, x_n(t)) = (\lambda_1(D), S_{in} - \lambda_1(D), 0, \cdots, 0)$$

It can be noted, as for the single-species chemostat model, that the variable "total mass":

$$m = s + \sum_{i=1}^{n} x_i$$

verifies the scalar differential equation:

$$\frac{dm}{dt} = D(S_{in} - m)$$

where $m = S_{in}$ is the unique steady-state, which is more generally asymptotically stable. Thus, any solution of the system [3.1] asymptotically joins the set:

$$I = \left\{ (s, x_1, \cdots, x_n) \in \mathbb{R}_+^{n+1} \, , \; s + \sum_{i=1}^{n} x_i = S_{in} \right\}$$

Before presenting a demonstration for proposition 3.2, we show a first result on the non-extinction of total biomass, of interest in itself.

LEMMA 3.1.– Let the variable "total biomass"

$$b(t) = \sum_{i=1}^{n} x_i(t)$$

Assume the condition:

$$\min_{i} \lambda_i(D) < S_{in}$$

is verified. Therefore, for any initial condition of the system [3.1] such that $b(0) > 0$, the function $t \mapsto b(t)$ is bounded from below by a strictly positive constant.

PROOF.– Observe that the dynamic of total biomass verifies the inequality:

$$\frac{db}{dt} = \sum_{i} \frac{dx_i}{dt} \geq \left(\min_{i} \mu_i(s) - D \right) b$$

Since the functions μ_i verify $\mu_i(S_{in}) > D$, there exist numbers $\epsilon > 0$ and $\eta > 0$ such that $\mu_i(\sigma) > D + \epsilon$ for all i and all $\sigma > S_{in} - \eta$. Moreover, since $t \mapsto m(t)$ converges to S_{in}, there exists a time $\bar{t}$ such that $m(t) > S_{in} - \eta/2$ for all $t > \bar{t}$. Thereby, the solution $t \mapsto b(t)$ verifies the following property:

$$b(t) < \frac{\eta}{2}, \; t > \bar{t} \; \Rightarrow \; \frac{db}{dt}(t) \geq (\min_{i} \mu_i(m(t) - \eta/2) - D)b(t) \geq \epsilon b(t)$$

We therefore deduce that the variable b cannot remain in the domain $\{b < \eta/2\}$ nor therein re-enter for times greater than $\bar{t}$. The variable b is thus bounded from below by the quantity $\eta/2$ after a finite time.   ∎

After a transient, we have:

$$s(t) + \sum_{i=1}^{n} x_i(t) \simeq S_{in}$$

Consequently, in the following two sections we consider the dynamics of the "reduced" system, in which we replace s by $S_{in} - \sum_j x_j$ in the expressions dx_i/dt:

$$\frac{dx_i}{dt} = \mu_i \left(S_{in} - \sum_{j=1}^{n} x_j \right) x_i - Dx_i \quad i = 1 \cdots n \qquad [3.6]$$

and since the variable s remains positive at all time, we only consider the solutions of this system inside the domain:

$$\Omega = \left\{ x = (x_1, \cdots, x_n) \in \mathbb{R}_+^n : \sum_{j=1}^{n} x_j \leq S_{in} \right\}$$

We show that for any initial condition where species 1 is present, the solutions of the reduced system asymptotically converge toward a unique steady-state where only species 1 is present. Although it may seem intuitive that the solutions of the non-reduced system also converge toward this equilibrium (for any initial condition where species 1 is present), its demonstration is based on the *theory of asymptotically autonomous dynamic systems*, which is explored in the appendices, but whose technical aspects fall beyond the scope of this book. A comprehensive proof making use of this argument is available in reference [BUT 85].

3.3.1. *A "graphical" proof for two species*

With regard to the case that involves two species, the system [3.6] is a system in the plane:

$$\begin{cases} \dfrac{dx_1}{dt} = (\mu_1(S_{in} - x_1 - x_2) - D)x_1 \\[2mm] \dfrac{dx_2}{dt} = (\mu_2(S_{in} - x_1 - x_2) - D)x_2 \end{cases} \qquad [3.7]$$

for which the isoclines are easily determined (see Appendix 1, section A1.4.2):

$$\frac{dx_1}{dt} = 0 \quad \Longleftrightarrow \quad x_1 = 0 \quad \text{or} \quad S_{in} - x_1 - x_2 = \lambda_1(D)$$

$$\frac{dx_1}{dt} = 0 \quad \Longleftrightarrow \quad x_2 = 0 \quad \text{or} \quad S_{in} - x_1 - x_2 = \lambda_2(D)$$

which consist of the axes and two parallel segments. Assuming that $\lambda_1(D) < \lambda_2(D) < S_{in}$, and by projecting the directions of the velocity vector over the isoclines, it can be seen that the trajectories can only reach the steady-state E_1, as soon as $x_1(0) > 0$ (see Figure 3.5). Analogously to the Lotka-Volterra competition model (see Appendix 1, section A1.4.2), when isoclines do not intersect one another outside the axes, it can be concluded that the steady-state E_1 is globally attractive in the positive quadrant. In Figure 3.5 (on the right), we have depicted a few solution trajectories of the reduced system.

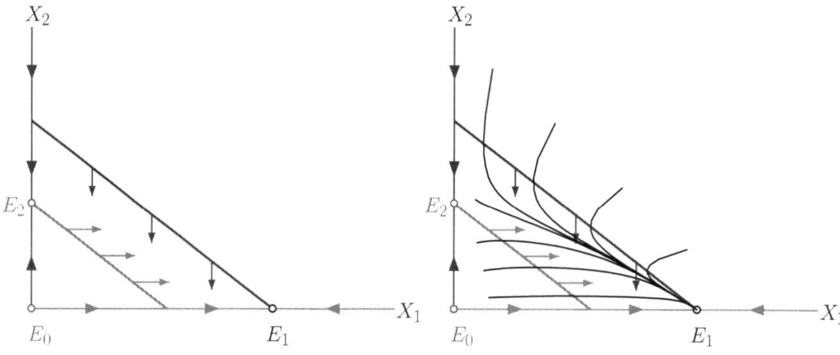

Figure 3.5. *Isoclines of the reduced model: $\dot{x}_1 = 0$ (in blue) and $\dot{x}_2 = 0$ (in green). Each intersection between isoclines determines a steady-state (red if unstable, black if stable). On the right: a few solution trajectories of the reduced system, which illustrate the global convergence toward the steady-state E_1. For a color version of this figure, see www.iste.co.uk/harmand/chemostat.zip*

3.3.2. *A proof for the general case*

The proof that we now present assumes the knowledge of ω-limits sets and of their properties (demonstration of lemma 3.5), which are recalled in the appendices.

To simplify the notation, we will denote in this section $\lambda_i(D)$ by λ_i.

Consider the set Δ defined by:

$$\Delta = \left\{ x \in \Omega : \sum_{j=1}^{n} x_j = S_{in} - \lambda_1 \right\}$$

An illustration of this set is given in Figure 3.6. This set splits the domain Ω into two regions denoted by $\mathcal{B}$ and $\mathcal{C}$:

$$\mathcal{B} = \left\{ x \in \Omega : \sum_{j=1}^{n} x_j < S_{in} - \lambda_1 \right\}, \quad \mathcal{C} = \left\{ x \in \Omega : \sum_{j=1}^{n} x_j > S_{in} - \lambda_1 \right\}$$

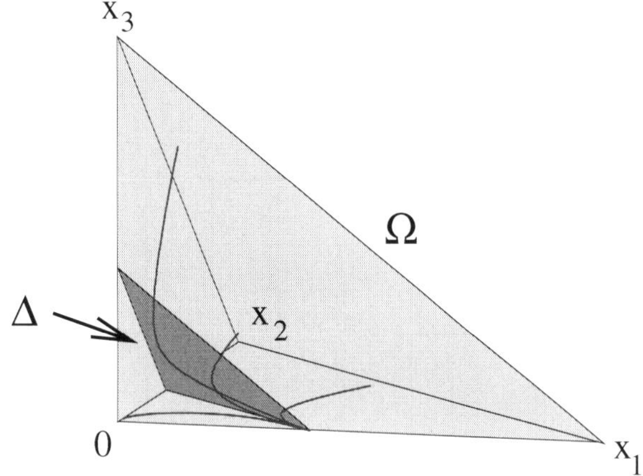

Figure 3.6. *Illustration of the set Δ included in the domain Ω for $n = 3$, with a few trajectories. For a color version of this figure, see www.iste.co.uk/harmand/chemostat.zip*

LEMMA 3.2.– Inside the region $\mathcal{C}$, all functions $t \mapsto x_i(t)$, $i = 1 \cdots n$ are strictly decreasing.

Demonstration. For any $i = 1 \cdots n$, since the functions μ_i are strictly increasing, we have:

$$\frac{dx_i}{dt} = \left(\mu_i \left(S_{in} - \sum_{j=1}^{n} x_j \right) - D \right) x_i$$

$$< (\mu_i(\lambda_1) - D) x_i \qquad \text{because} \quad S_{in} - \sum_{j=1}^{n} x_j < \lambda_1 \text{ in } \mathcal{C}$$

$$< (\mu_i(\lambda_i) - D) x_i \qquad \text{because} \quad \lambda_1 < \lambda_i.$$

Since $\mu_i(\lambda_i) = D$, we have thus showed that $dx_i/dt < 0$ within $\mathcal{C}$.

Consider an initial condition in $\mathcal{C}$. Two cases can occur: either the solution always remains inside $\mathcal{C}$, or it leaves $\mathcal{C}$ to enter $\mathcal{B}$. We are going to show (see lemma 3.3) that in the first case, the solution necessarily converges toward the steady-state:

$$E_1 = (\lambda_1, S_{in} - \lambda_1, 0, \cdots, 0)$$

and in the second case, the solution will no longer leave $\mathcal{B}$. As soon as it is inside $\mathcal{B}$, the solution also tends toward E_1 (see lemma 3.5).

Let us begin by examining the first case.

LEMMA 3.3.– If the solution remains in $\mathcal{C}$ for any $t > 0$, it then converges to E_1.

Demonstration. Given that the functions $t \mapsto x_i(t)$, $i = 1 \cdots n$ are strictly decreasing and positive, the limits:

$$\lim_{t \to +\infty} x_i(t) = c_i, \qquad i = 1 \cdots n$$

do exist. It can be thereof deduced (see the appendices) that:

$$c = (c_1, \cdots, c_n)$$

is a steady-state of [3.6] and that this steady-state belongs to $\mathcal{C}$, or to its boundary. Since E_1 is the only steady-state verifying this property, we can therefore infer that $c = E_1$, in other words that all the solutions that remain in $\mathcal{C}$ converge toward E_1.

Now let us examine the second case.

LEMMA 3.4.– If the solution enters $\mathcal{B}$ then it can no longer leave.

Demonstration. For all $x \in \Delta \setminus E_1$, given that the functions μ_i are strictly increasing, we can write:

$$\begin{aligned}
\frac{d}{dt} \sum_{i=1}^{n} x_i &= \sum_{i=1}^{n} \left(\mu_i \left(S_{in} - \sum_{j=1}^{n} x_j \right) - D \right) x_i \\
&= \sum_{i=1}^{n} \left(\mu_i(\lambda_1) - D \right) x_i \quad \text{because } S_{in} - \sum_{j=1}^{n} x_j = \lambda_1 \text{ in } \Delta \\
&= \sum_{i=2}^{n} \left(\mu_i(\lambda_1) - D \right) x_i \quad \text{because } \mu_1(\lambda_1) = D \\
&< \sum_{i=2}^{n} \left(\mu_i(\lambda_i) - D \right) x_i \quad \text{because } \lambda_1 < \lambda_i \text{ (and } \mu_i(\lambda_1) < \mu_i(\lambda_i))
\end{aligned}$$

Since $\mu_i\left(\lambda_i\right) = D$ for all i, we have thus showed that in Δ, we have:

$$\frac{d}{dt}\sum_{i=1}^{n} x_i < 0$$

Therefore, Δ which is defined by $\sum_{j=1}^{n} x_j = S_{in} - \lambda_1$ can only be crossed by a solution when shifting from $\mathcal{C}$, which is defined by $\sum_{j=1}^{n} x_j > S_{in} - \lambda_1$ to $\mathcal{B}$, which is defined by $\sum_{j=1}^{n} x_j < S_{in} - \lambda_1$.

Let us now study the behavior of a solution originating from $\mathcal{B}$.

LEMMA 3.5.– Any solution originating from a point of $\mathcal{B}$ tends toward E_1.

Demonstration. Since the function μ_1 is strictly increasing and that a solution coming from $\mathcal{B}$ cannot leave it, it follows that:

$$\frac{dx_1}{dt} = \left(\mu_1\left(S_{in} - \sum_{j=1}^{n} x_j\right) - D\right) x_1$$

$$> \left(\mu_1\left(\lambda_1\right) - D\right) x_1 \qquad \text{because} \quad S_{in} - \sum_{j=1}^{n} x_j > \lambda_1 \text{ in } \mathcal{B}$$

Given that $\mu_1(\lambda_1) = D$, it thus has been proved that $dx_1/dt > 0$ in $\mathcal{B}$. Consequently, the function $t \mapsto x_1(t)$ is strictly increasing and upper bounded by $S_{in} - \lambda_1$ (since the solution cannot leave $\mathcal{B}$). It can thereof be deduced that the limit:

$$\lim_{t \to +\infty} x_1(t) = c_1 > 0 \qquad [3.8]$$

exists. The limit is strictly positive because $x_{1(0)} > 0$.

The technique that we adopt hereafter for the proof follows the approach of LaSalle's principle of invariance. We denote by ω the limit set of the solution $x(t)$. It is non-empty because the solution remains in $\mathcal{B}$. It is included in $\mathcal{B}$ or in its boundary, that is in $\mathcal{B} \cup \Delta$. It is also invariant (this is a general property of limit sets, see appendices). From [3.8], it can be derived that:

$$\omega \subset \{x \in \mathcal{B} \cup \Delta : x_1 = c_1\} \qquad [3.9]$$

In effect, for any point of $y \in \omega$, there exits a sequence t_n tending to infinity, such that:

$$y = \lim_{n \to +\infty} x(t_n)$$

Consequently, we have $y_1 = c_1$.

Moreover, provided that x_1 is constant in ω (and that ω is invariant), we thereof deduce that

$$\omega \subset \left\{ x \in \mathcal{B} \cup \Delta : \frac{dx_1}{dt} = 0 \right\} = \{ x \in \mathcal{B} \cup \Delta : x_1 = 0 \text{ or } x \in \Delta \} \qquad [3.10]$$

The two inclusions [3.9] and [3.10] show that $\omega \subset \Delta$, because $c_1 > 0$. Since $\{E_1\}$ is the only invariant set included in Δ, we deduce that $\omega = \{E_1\}$. As a consequence, all solutions issuing from $\mathcal{B}$ tend toward E_1. ∎

Finally, using lemmas 3.3 and 3.5, we have demonstrated that any solution with a strictly positive initial condition tends toward E_1, which concludes the demonstration of proposition 3.2.

EXERCISE 3.3.– In this exercise, the comprehensive system [3.1] is considered in $\mathbb{R}^{n+1}$, but with initial conditions within the set:

$$\Omega = \left\{ (s, x) \in \mathbb{R}_+^{n+1} : s + \sum_{i=1}^{n} x_i \leq S_{in} \right\}$$

It is assumed that $\lambda_1 < S_{in}$:

1) show that for any initial condition in Ω, the solution remains in Ω for any positive time;

2) let the subset $\mathcal{C} = \{ (s, x) \in \Omega : s > \lambda_1 \}$. Show that in the set $\mathcal{C}$, the function $t \mapsto x_1(t)$ is strictly increasing, when $x_1(0) > 0$;

3) show that any solution with $x_1(0) > 0$ that remains in $\mathcal{C}$ necessarily converges to the steady-state E_1;

4) let the subset $\mathcal{B} = \{ (s, x) \in \Omega : s < \lambda_1 \}$. Show that any solution with $x_1(0) > 0$ that remains in $\mathcal{B}$ necessarily converges toward the steady-state E_1;

5) show that if a solution enters $\mathcal{B}$, it cannot subsequently leave it;

6) show then that for an initial condition in Ω such that $x_1(0) > 0$, the solution converges toward E_1.

3.4. The case of non-monotonic growth functions

In this section, we consider that the growth functions μ_i of the model [3.1] can be non-monotonic.

3.4.1. *Growth set*

For each species i, it is helpful to introduce the *growth set* defined as follows:

$$\Lambda_i(D) = \{s > 0 \text{ such that } \mu_i(s) > D\}$$

Therefore, for a non-empty set $\Lambda_i(D)$, the break-even concentration previously introduced is $\lambda_i(D) = \inf \Lambda_i(D)$. As in the previous chapter, we consider that non-monotonic growth functions are strictly increasing and then strictly decreasing, which is tantamount to requiring that sets $\Lambda_i(D)$ (when they are non-empty) are intervals:

$$\Lambda_i(D) =]\lambda_i(D), \bar{\lambda}_i(D)[$$

where $\bar{\lambda}_i(D) = \sup \Lambda_i(D)$ is a "maximum" break-even concentration, which can be equal to $+\infty$ (in particular, for a monotonic function μ_i such that $\Lambda_i(D) \neq \emptyset$, it follows that $\bar{\lambda}_i(D) = +\infty$).

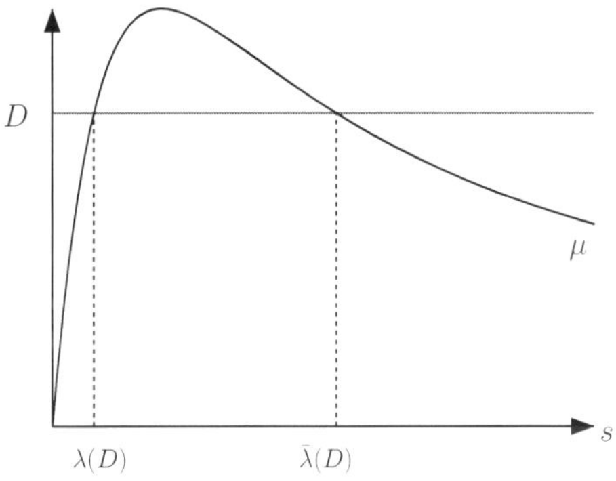

Figure 3.7. *Break-even concentration* $\lambda(D), \bar{\lambda}(D)$

For a Haldane expression:

$$\mu(s) = \frac{\mu_0\, s}{K_S + s + s^2/K_I}$$

the set $\Lambda(D)$ is non-empty under the condition $D < \max_s \mu(s)$.

EXERCISE 3.4.–

1) Show that for a Haldane function, we have:

$$\max_s \mu(s) = \mu(\sqrt{K_S K_I}) = \frac{\mu_0}{1 + 2\sqrt{K_S/K_I}} \tag{3.11}$$

2) Show that with the condition $D < \mu_0/(1 + 2\sqrt{K_S/K_I})$, break-even concentrations are given by the following formulae:

$$\lambda(D) = K_I \frac{\mu_0/D - 1 - \sqrt{(\mu_0/D - 1)^2 - 4K_S K_I}}{2}$$

$$\bar{\lambda}(D) = K_I \frac{\mu_0/D - 1 + \sqrt{(\mu_0/D - 1)^2 - 4K_S K_I}}{2}$$

[3.12]

3.4.2. *Study of steady-states*

As justified above, we consider that situations such that values $\lambda_i(D)$ or $\bar{\lambda}_i(D)$ coincide with values $\lambda_j(D)$ or $\bar{\lambda}_j(D)$ or S_{in} are exceptional. Thus, except for those situations, competitive exclusion at steady-state remains verified. The steady-states of the system are therefore the washout E_0 and the steady-states for which $\mu_i(s_i^\star) = D$. Now, there are two ways to achieve the equality $\mu_i(s_i^\star) = D$: the steady-states E_i when $\lambda_i(D) < S_{in}$, but also steady-states $\bar{E}_i$ where variables s and x_i at steady-state are $\bar{s}_i^\star = \bar{\lambda}_i(D)$ and $\bar{x}_i^\star = S_{in} - \bar{s}_i^\star$, when $\bar{\lambda}_i(D) < S_{in}$.

By recalling the stability study of section 3.1.3, it is verified that the stability condition of the washout steady-state E_0 is unchanged: E_0 is (locally exponentially) stable when $\lambda_i(D) > S_{in}$ for all i. A steady-state $\bar{E}_i$ (when it exists) cannot be stable, since, as with the monospecific chemostat model, $-\mu_i'(\bar{s}_i)\bar{x}_i$ is a strictly positive eigenvalue. Finally, a steady-state E_i (where it exists) is stable when $\mu_j(s_i^\star) < D$ for all $j \neq i$. It is thus once again found that the steady-state $E_{i^\star}$ corresponding to the smallest break-even concentration is stable. However, unlike the case in which all functions μ_i are monotonic, there is a possibility to achieve stability for another steady-state E_i for $i \neq i^\star$ when $s_i^\star = \lambda_i(D)$ does not belong to any set $\Lambda(j)(D)$ for $j \neq i$. This situation amounts to requiring that the union of all $\Lambda(i)(D)$ is not an interval but a union of at least two disjoint intervals. This is formalized in the next section and illustrated in section 3.4.5.

3.4.3. *Competitive exclusion*

We have, similarly to the case of monotonic functions, proposition 3.3.

PROPOSITION 3.3.– Given the set:

$$\mathcal{E}(S_{in}, D) = \bigcup_{i=1}^{n} \Lambda_i(D) \cap \,]0, S_{in}[$$

1) the washout steady-state is (locally exponentially) stable when $S_{in} \notin \mathcal{E}(S_{in}, D)$;

2) when $\mathcal{E}(S_{in}, D)$ is not empty, it is written as a union of m disjoint intervals:

$$\mathcal{E}(S_{in}, D) = \bigcup_{k=1}^{m} \,]\lambda_{i_k}, \min(\bar{\lambda}_{j_k}, S_{in})[$$

where m is an integer between 1 and n. Then the system [3.1] admits exactly m steady-states E_{i_k} $(k = 1 \cdots m)$ (locally exponentially) stable.

It should be noted that when the system admits several stable steady-states, the steady-state reached by the trajectories of the system depends of the initial condition. This is illustrated with an example in section 3.4.5.

EXERCISE 3.5.– When the growth functions μ_i are monotonic in the interval $[0, S_{in}]$, show that this proposition exactly gives the result of proposition 3.1 stated in section 3.2.

In the next section, we describe the set of possible cases for two species. For the general case with more than two species, the combinatorics of the possibilities of intersection between the sets $\Lambda_i(D)$ and their positioning with respect to the value S_{in} may prove very large. Our objective here is not to make a comprehensive classification of the possible cases but to provide a methodology for obtaining the number of stable steady-states, which consists in determining:

– the number of disjoint intervals of the set:

$$E = \bigcup_i \Lambda_i(D)$$

– the positioning of the parameter S_{in} with respect to this set, that is, the number of interval(s) of the set $E \cap [0, S_{in}]$, which indicates the number of stable steady-states outside washout;

the membership of S_{in} to $E \cap [0, S_{in}]$, which provides information on the stability of the washout steady-state.

3.4.4. *Competition between two species*

The determination of the possible steady-states and their stability (following the conditions of proposition 3.3) is tantamount to studying the set:

$$\mathcal{E} = (\Lambda_1(D) \cup \Lambda_2(D)) \cap [0, S_{in}]$$

according to the operating conditions (D, S_{in}). The five possible situations (excluding "exceptional" situations) are presented in Table 3.2.

Cases 3 to 5 require that at least one of the two functions be non-monotonic. Only case 5 is specific to two non-monotonic functions.

As previously, this table only shows the results of a local study of steady-state. It can be shown that the global behavior of the solutions of the system always consists in asympotically converging toward a steady-state. This corresponds to a result rather technical to demonstrate, which falls outside of the scope of this book. For this reason, we refer the reader to the bibliographic notes of section 3.5.

3.4.5. *Illustration and effect of a "bio-augmentation"*

We illustrate this result with a configuration including two species in which only the first presents non-monotonic kinetics, according to the following Haldane and Monod expressions:

$$\mu_1(s) = \frac{s}{1 + s + s^2}, \quad \mu_2(s) = \frac{(1/2)s}{2 + s}$$

The formula [3.11] gives us $\max_s \mu_1(s) = 1/3$. Thereby, for values of D lower than $1/3$, the sets $\Lambda_1(D)$ and $\Lambda_2(D)$ are not empty. We consider the following two situations:

– $D = 1/4$: break-even concentrations given by formulae [3.12] and [3.2] are:

$$\lambda_1(1/4) = \frac{3 - \sqrt{5}}{2}, \quad \bar{\lambda}_1(1/4) = \frac{3 + \sqrt{5}}{2}, \quad \lambda_2(1/4) = 2$$

Since $\lambda_2(1/4) \in]\lambda_1(1/4), \bar{\lambda}_1(1/4)[$, only species 1 can remain in a stable way at steady-state. This therefore corresponds to Case 2 in Table 3.2;

– $D = 3/10$: we get:

$$\lambda_1(3/10) = \frac{7/3 - \sqrt{13/9}}{2}, \quad \bar{\lambda}_1(3/10) = \frac{7/3 + \sqrt{13/9}}{2}, \quad \lambda_2(3/10) = 3$$

This time, it follows that $\lambda_2(3/10) > \bar{\lambda}_1(3/10)$. There is therefore possibility of having species 1 or species 2 at a stable steady-state. This is thus Case 4 of Table 3.2.

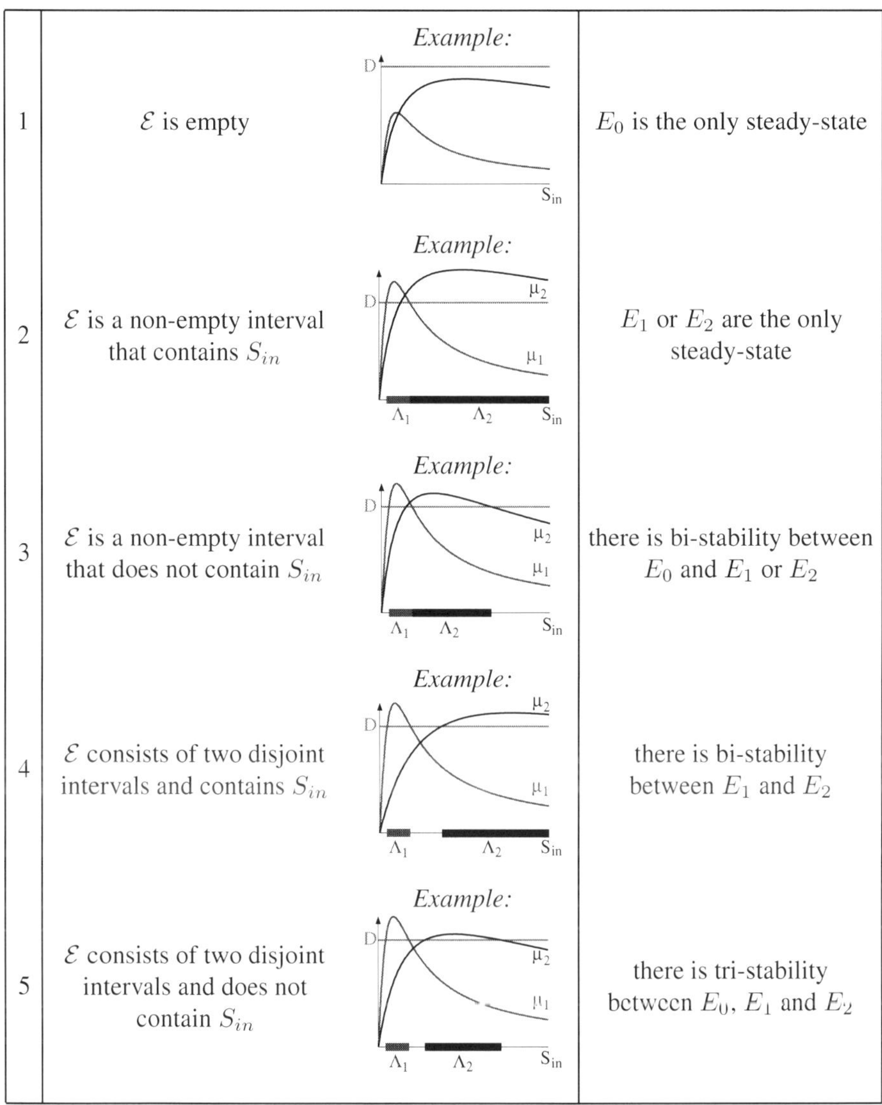

1	$\mathcal{E}$ is empty	*Example:*	E_0 is the only steady-state
2	$\mathcal{E}$ is a non-empty interval that contains S_{in}	*Example:*	E_1 or E_2 are the only steady-state
3	$\mathcal{E}$ is a non-empty interval that does not contain S_{in}	*Example:*	there is bi-stability between E_0 and E_1 or E_2
4	$\mathcal{E}$ consists of two disjoint intervals and contains S_{in}	*Example:*	there is bi-stability between E_1 and E_2
5	$\mathcal{E}$ consists of two disjoint intervals and does not contain S_{in}	*Example:*	there is tri-stability between E_0, E_1 and E_2

Table 3.2. *The five possible outcomes of the competition at steady-state between two species whose growth curves are not necessarily monotonic. For a color version of this table, see www.iste.co.uk/harmand/chemostat.zip*

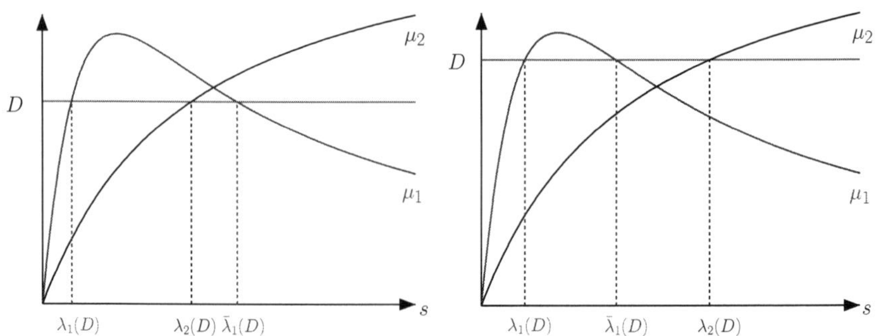

Figure 3.8. *Example where according to the value of D, the sets $\Lambda_1(D)$ and $\Lambda_2(D)$ may not be separated (on the left for $D = 1/4$) or be separated (on the right for $D = 3/10$). For a color version of this figure, see www.iste.co.uk/harmand/ chemostat.zip*

Figure 3.8 shows the relative positions of numbers $\lambda_1(D)$, $\bar{\lambda}_1(D)$, $\lambda_2(D)$ for these two situations, over the interval $[0, 5]$. As in section 3.3, we consider the reduced model [3.7] in the plane, for which the isoclines are represented in Figure 3.9. Figure 3.10 illustrates how the positioning of the isocline $\dot{x}_2 = 0$ modifies the phase portrait and creates two basins of attraction. It should be noted that in these two situations (and for the value of S_{in} equal to 5), the washout steady-state is stable for the monospecific model in the absence of species 2, since we have $\mu_1(S_{in}) = 5/31 < D$. Thus, the particular trajectories that remain on the axis $x_2 = 0$ can join E_0. Nevertheless, the washout steady-state E_0 is unstable when considering the model with both species, since we have $\mu_2(S_{in}) = 5/14 > D$. In this example, it can be seen that for a chemostat with a single species which presents a bi-stability, as for species 1, a possibility for systematically avoiding washout is to add a species whose characteristics are similar to those of species 2. Indeed:

– in the first situation ($D = 1/4$), species 2 is useful in avoiding washout, but cannot remain alone at steady-state, unlike species 1;

– in the second situation ($D = 3/10$), species 2 is always useful in avoiding washout, but cannot remain alone at steady-state to the detriment of species 1.

Thereof, it can be derived that if we wish to culture in a chemostat at steady-state a species presenting inhibition in operating conditions such that $S_{in} \notin \Lambda(D)$, it is benefitial to proceed to a "bio-augmentation" by adding another species that does not present any inhibition and such that its break-even concentration belongs to $\Lambda(D)$. The biological interpretation is that this additional species will enable, in a first stage, the substrate concentration to be decreased to values small enough for which the species of interest can grow without suffering the effects of inhibition.

As in section 3.2, we summarize the set of the possibilities of stable steady-states for the two growth curves using the operating diagram given in Figure 3.11.

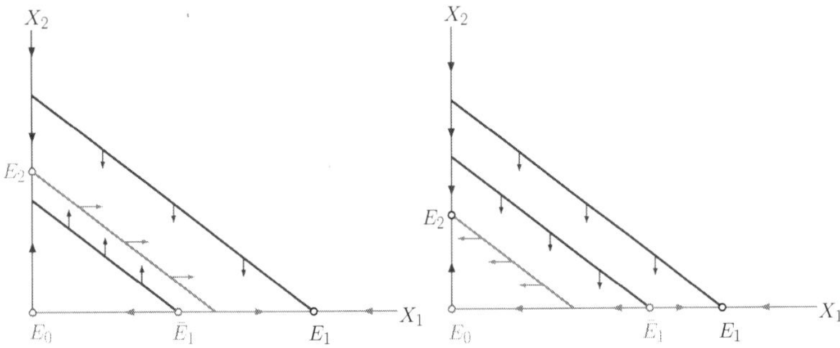

Figure 3.9. *Isoclines, in blue:* $\dot{x}_1 = 0$, *in green:* $\dot{x}_2 = 0$. *Each intersection between isoclines determines a steady-state (red if unstable, black if stable). For a color version of this figure, see www.iste.co.uk/harmand/chemostat.zip*

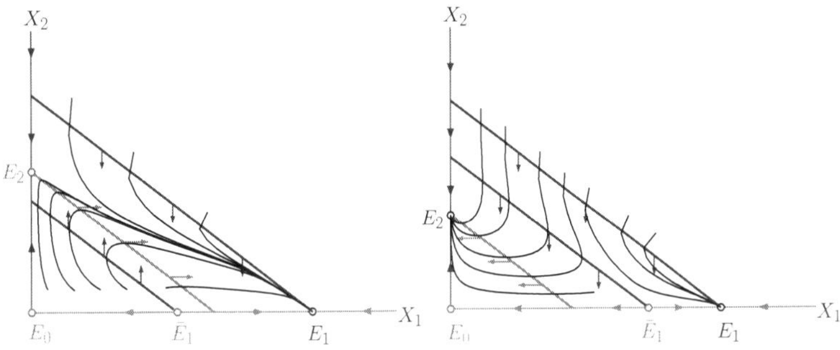

Figure 3.10. *A few trajectories of the reduced system. For a color version of this figure, see www.iste.co.uk/harmand/chemostat.zip*

EXERCISE 3.6.–

1) show that the graphs of a Monod and of a Haldane function cannot have more than two intersections outside of 0;

2) determine the operating diagram corresponding to functions:

$$\mu_1(s) = \frac{s}{1 + s + s^2}, \quad \mu_2(s) = \frac{3s}{1 + 10s}$$

EXERCISE 3.7.– Show that for operating parameters $D = 3/10$ at $S_{in} = 5$, the two-species chemostat model whose growth functions are:

$$\mu_1(s) = \frac{2s}{1 + s + 10s^2/2}, \quad \mu_2(s) = \frac{6s}{10(1 + s + s^2/5)}$$

presents a three-stability.

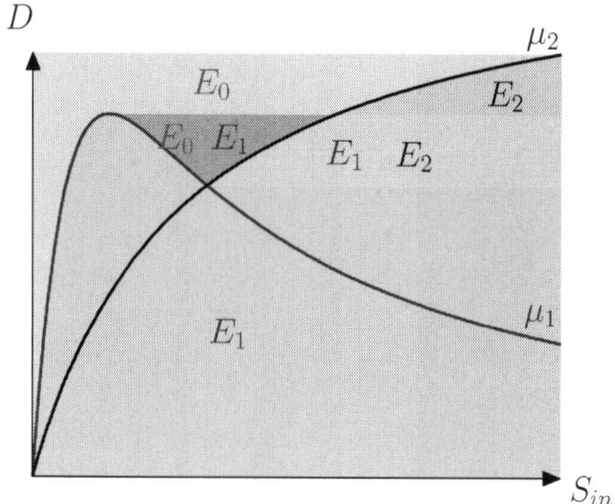

Figure 3.11. *Operating diagram of the stability of steady-states E_0, E_1 and E_2. For a color version of this figure, see www.iste.co.uk/harmand/chemostat.zip*

3.5. Bibliographic notes

As early as the 1930s, the Russian botanist Gause conducted experiments on the growth of yeasts and paramecia in *mixed* cultures [GAU 32], and reported that the most competitive species systematically eliminated the other in a few days (see Figure 3.12). In his book on "the struggle for existence" [GAU 03], he showed that competitive exclusion had a more universal scope when living species were in competition for a shared resource. In fundamental ecology and in population biology, an "exclusion principle" is commonly taught also known as "Gause's law", which states that two similar species evolving in the same medium and requiring the same resources cannot coexist at steady-state in this medium (one of these two species will always have a slight advantage over the second, and this will lead in the long term to the extinction of the species the less well adapted to the ecosystem in question).

This statement, which has been extended to more than one limiting resource (stating that at a stable steady-state it is not possible to observe more than n competitors) has become very popular in theoretical ecology in the 1960s: the *principle of competitive exclusion* applies to all kinds of ecosystems (in other words, not just those concerning microorganisms), since there are *consumers* and *resources*.

However, it was not until the 1970s that the first statement of a mathematical theorem was found in the literature, along with its proof [HSU 77], for the specific model of the chemostat which is of interest to us here. The *competitive exclusion principle* generally refer to a mathematical result that establishes conditions under which almost all solutions converge toward a steady-state of the system having at most one species.

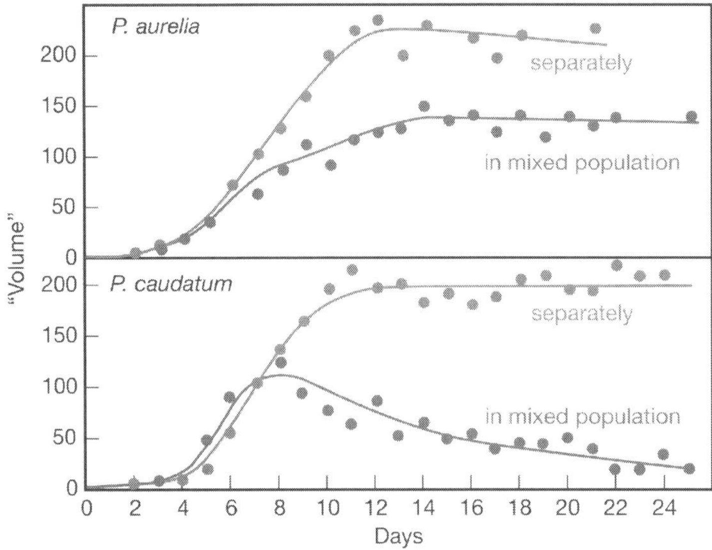

Figure 3.12. *Records of Gause experiments (image from [GAU 03])*

In the 1980s, Hansell and Hubbell's laboratory experiments allowed for the observation of the competitive exclusion in chemostats, such as predicted by the mathematical theory. In these experiments, the addition of an auxiliary substance which acts on the growth curve of only one of the two species, has also made it possible to obtain a situation corresponding to a small value of number ν (see section 3.2.3) and to observe the preservation of both species over relatively long periods (see Figure 3.13).

Under the conditions of proposition 3.1 (which gives only one condition for the local stability of the steady-state $E_i^\star$), the proof of the overall convergence of any trajectory toward steady-state $E_i^\star$ (when it exists) when $x_i^\star(0)$ is not zero, has a long history in the literature of bio-mathematics. Hsu, Hubbell and Waltman have proposed a first proof in 1977 for Monod functions and identical removal rates [HSU 77]. Hsu generalized this result in 1978 for different removal rates [HSU 78]. These two contributions utilize an explicit Lyapunov function to demonstrate the overall convergence (see Appendix 1 for the concept of the Lyapunov function). In 1980, Armstrong and McGehee have given a simple proof for any monotonic growth functions and identical removal rates [ARM 80], but for initial conditions belonging to the invariant set:

$$s + \sum_i x_i = S_{in}$$

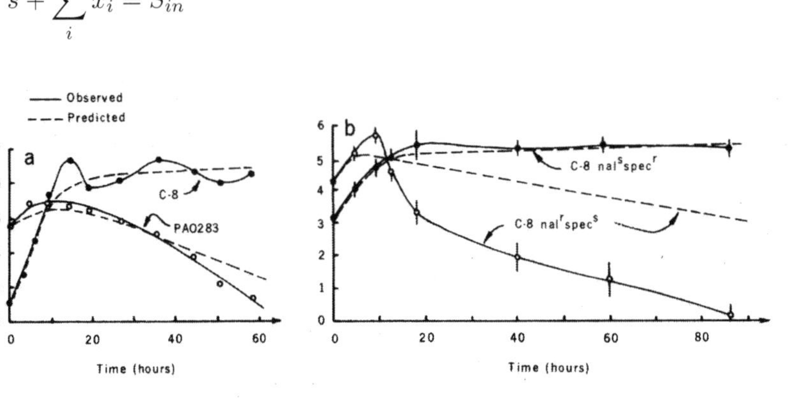

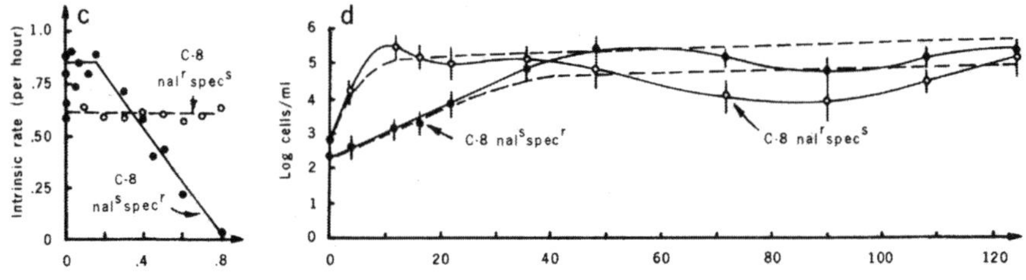

Figure 3.13. *Records of experiments by Hansen and Hubbell (image from [HAN 80])*

In 1985, Butler and Wolkowicz proposed a proof for any monotonic growth function, but with identical removal rates [BUT 85]. One of the difficulties originates from the fact that the graphs of any growth function can intersect one another at several points. Finally, it was in 1992 that Wolkowicz and Lu proposed a proof, based on a Lyapunov function, for growth functions more general than Monod functions (but under additional technical assumptions) and different removal rates [SAR 13, WOL 92]. These proofs are beyond the scope of this book, because of their technical nature. Several refinements have followed, which we do not detail here, alleging these technical assumptions. However, the proof of overall stability for any monotonic growth function and different removal rates remains an open mathematical problem [DEL 03].

It was around the 1970s that Andrews showed that known inhibition phenomena for enzymes can also apply to the growth of certain microorganisms under strong concentration of limiting substrate, and suggested using Haldane's expression to model specific growth curves [AND 68]. The competition between two species in which one or both exhibit inhibition has been studied for example in references [ARI 77, DER 75]. The mathematical results of competitive exclusion have been revisited for this type of functions, and proofs of the convergence of almost any trajectory of the system [3.1] toward one of the stable steady-state are available with references [BUT 85, LI 98, RAP 08, WOL 92]. These results do not necessarily require that growth curves accurately follow Haldane's expression but to be increasing and then decreasing, as we have assumed in section 3.4. The overall stabilization of the model with a single bi-stable species, through "bio-augmentation" of a "well-chosen" species is studied and described in references [HAR 08, RAP 08, WAD 16].

4

Competition: the Density-Dependent Model

4.1. Chapter orientation

This chapter contrasts with Chapters 2 and 3 whose messages were fairly simple:

– the size of the population of an isolated species cultivated in a chemostat reaches a stable steady-state, which can be 0 when the dilution D is too significant (Chapter 2), or in the case of a growth rate of the "Haldane" type;

– wihtin a chemostat several species cannot exist simultaneously at steady-state. This is the competitive exclusion theorem as stated in Chapter 3.

Moreover, readers should have surely noticed that these mathematical models seemed quite distant in their conclusions from the complexity of reality but they should have realized that, and with reason, this simplicity was the result of pedagogical imperatives. In this chapter, we are going to address the modeling of phenomena more complex than the simple fact of consuming substrate: *direct interference* between bacteria.

In our minimal model, bacteria interfere because the substrate consumed by some bacteria will not be able to be used by another. However, we had noticed in our extension of the model to density-dependent growth rate that bacteria could "interfere" and that a density which is too large could limit their access to substrate, or even produce toxins having a negative effect. Whereas the density-dependent model that we had built to account for this issue kept all the properties of the minimal model, the same does not apply from the moment there are more than two existing species. We will see that if we consider direct interference, the conclusions of Chapter 3 are dramatically altered. The coexistence which was exceptional will become very common.

Before proceeding, the terminology about the concept of competition should be somewhat clarified. Consider the very general model (Kolmogorov model) describing the interaction of n biological species, but that does not explicitly consider any substrate being consumed:

$$\frac{dx_i}{dt} = f_i(x_1, \cdots, x_i, \cdots, x_n)x_i \qquad i = 1 \cdots n \qquad [4.1]$$

In theoretical ecology, it is said that species j exerts a competition pressure on species i if *the increase of density of x_j decreases the growth rate of species i* or, which amounts to the same, if:

$$\partial_{x_j} f_i(x_1, \cdots, x_i, \cdots, x_n) < 0$$

If $i \neq j$, it is then referred to as interspecific competition, otherwise this is intraspecific competition. It is important to notice that this definition raises a problem for a model where the f_i depend on environmental variables such as, for example, a substrate because this would be equivalent to state that in the model:

$$\begin{cases} \dfrac{ds}{dt} &= D(S_{in} - s) - \sum_{i=1}^{n} \mu_i(s)x_i \\ \dfrac{dx_i}{dt} &= (\mu_i(s) - D)x_i \quad i = 1 \cdots n \end{cases} \qquad [4.2]$$

the species are not in competition since all the $\partial_{x_j}(\mu_i(s) - D)$ are equal to zero! Nevertheless, we need to distinguish this competition for the competition for substrate within the meaning of the Kolmogorov model, which we will call *(direct) intraspecific competition* and *(direct) interspecific competition*.

The purpose of this chapter is thus to establish a number of properties for the model:

$$\begin{cases} \dfrac{ds}{dt} &= D(S_{in} - s) - \sum_{i=1}^{n} \mu_i(s,x)x_i \\ \dfrac{dx_i}{dt} &= (\mu_i(s,x) - D_{x_i})x_i \quad i = 1 \cdots n \end{cases} \qquad [4.3]$$

with:

$$x = (x_1, \cdots, x_n)$$

representing n different species consuming a single substrate in a perfectly-mixed chemostat. Concentrations are represented by s and x_i. All the yield coefficients *assumed as constant* have been reduced to 1 through the choice of suitable unit values. We also assume, *which considerably facilitates the study*, that all D_{x_i} are equal to D. In effect, as already done in previous chapters, if we set that: $z = s + \sum_{i=1}^{n} x_i$ $D_{x_i} = D$, by adding up all the equations of [4.3] it yields that $\frac{dz}{dt} = D(S_{in} - z)$. Therefore, we can consider that after a transient "it is as if" z was equal to S_{in} (see section A1.3.3 concerning "it is as if") and thus reduce the system [4.3] of $n + 1$ equations to the system of n equations:

$$\left\{ \frac{dx_i}{dt} = \left(\mu_i \left(S_{in} - \sum_{j=1}^{n} x_j, x \right) - D \right) x_i \quad i = 1, \cdots, n \right. \qquad [4.4]$$

NOTATION 4.1.– To simplify the notations, we will use throughout this chapter the following ones:

$$\partial_j \mu_i \equiv \partial_{x_j} \mu_i$$

We make the following assumptions:

HYPOTHESIS 4.1.–

1) for any i, we have $\mu_i(s, x) \geq 0$ and $\mu_i(0, x) = 0$;

2) for any i, we have $\partial_s \mu_i(s, x) > 0$;

3) for any i and any j, we have $\partial_j \mu_i(s, x) \leq 0$;

4) for any i, the function μ_i is bounded.

These assumptions are to be interpreted in the following way.

– The functions:

$$s \mapsto \mu_i(s, 0)$$

represent the limits of growth rates for infinitesimal population densities, therefore when there is virtually no interference between cells of various species; we will call them, as in section 2.3.3, *intrinsic growth rates*.

– Hypothesis 2 means that in this chapter, we will not consider inhibition phenomena;

– Hypothesis 3 means that species j exerts a (direct) competition pressure (possibly equal to zero) on species i but also that we exclude any form of mutualism, whether within species or between different species. Thus:

- $\partial_i \mu_i$ represents the intensity of the (direct) intraspecific competition;

- $\partial_j \mu_i$ $i \neq j$ represents the intensity of (direct) interspecific competition.

First, we are going to consider the case of two species for which we will be able to complete a comprehensive discussion. This is mainly due to the fact that the system is reduced to a differential system in the plane for which we can use simple and effective tools. The main message from this analysis will be that, contrary to the assertion of the competitive exclusion theorem of the previous chapter, the *coexistence of two species competing for a single substrate is perfectly possible*: it suffices that (direct) intraspecific competition prevails (within a sense that will be specified) over (direct) interspecific competition.

We will then discuss the case involving any number of species but in the particular context of *pure-intraspecific* competition, namely the case in which μ_i depends only on s and x_i excluding all other species x_j or, if one prefers: $i \neq j \implies \partial_j \mu_i \equiv 0$. Under this assumption, we will demonstrate the existence of a *coexistence steady-state* in which several species, or even all species are present, which completely contrasts with the competitive exclusion theorem of Chapter 3. Intuitively, there is a simple explanation: competition within species i limits the growth of population i, which leaves substrate available for other less performing species. The hypothesis of *pure-intraspecific* competition is most particular, but as we will see, not totally unrealistic.

We will close this chapter by a study (in simulation) focusing on three species in the presence of increasingly strong (direct) interspecific competition. We will see that the system shifts from a coexistence steady-state to a *dynamic coexistence* where species densities periodically oscillate and eventually achieve a form of exclusion different from those examined so far.

4.2. Two-species competition

We consider the model [4.3] when $n = 2$, let:

$$\begin{cases} \dfrac{ds}{dt} &= D(S_{in} - s) - \big(\mu_1(s, x_1, x_2)x_1 + \mu_2(s, x_1, x_2)x_2\big) \\[2mm] \dfrac{dx_1}{dt} &= (\mu_1(s, x_1, x_2) - D)x_1 \\[2mm] \dfrac{dx_2}{dt} &= (\mu_2(s, x_1, x_1) - D)x_2 \end{cases} \qquad [4.5]$$

where μ_1 and μ_2 satisfy the assumptions 4.1. If we consider the variable $z = s + x_1 + x_2$, we immediately get $\frac{dz}{dt} = D(S_{in} - z)$ from which it follows that trajectories are bounded and that the set:

$$I = \{(s, x_1, x_2) : s + x_1 + x_2 = S_{in}, \quad x_1 \geq 0, \quad x_2 \geq 0\}$$

is invariant and attractive. We focus on this set only and from now on we therefore study:

$$\begin{cases} \dfrac{dx_1}{dt} &= (\mu_1(S_{in} - (x_1 + x_2), x_1, x_2) - D)x_1 \\ \dfrac{dx_2}{dt} &= (\mu_2(S_{in} - (x_1 + x_2), x_1, x_2) - D)x_2 \end{cases} \qquad [4.6]$$

in the set $\{(x_1, x_2) : x_1 \geq 0, \quad x_2 \geq 0, \quad x_1 + x_2 \leq S_{in}\}$. We are not diminishing the generality because, after a transient, it can be considered that from a practical perspective the solution is effectively in this set.

4.2.1. *Behavior of an isolated species*

When the other species is absent, species i behaves according to the model:

$$\begin{cases} \dfrac{ds}{dt} &= D(S_{in} - s) - \mu_1(s, x_1, 0)x_1 \\ \dfrac{dx_1}{dt} &= \big(\mu_1(s, x_1, 0) - D\big)x_1 \end{cases} \qquad [4.7]$$

for species 1 and:

$$\begin{cases} \dfrac{ds}{dt} &= D(S_{in} - s) - \mu_2(s, 0, x_2)x_2 \\ \dfrac{dx_2}{dt} &= \big(\mu_2(s, 0, x_2) - D\big)x_2 \end{cases} \qquad [4.8]$$

for species 2. In both cases, this corresponds to the model studied in section 2.3.3 for which we have defined the characteristic at equilibrium. Therefore, we have two characteristics at equilibrium defined for $i = 1, 2$ by:

$$\begin{cases} s \leq \lambda_1 &\implies \psi_1(s) = 0 \\ s \geq \lambda_1 &\implies \mu_1(s, \psi_1(s), 0) = D \\[6pt] s \leq \lambda_2 &\implies \psi_2(s) = 0 \\ s \geq \lambda_2 &\implies \mu_2(s, 0, \psi_2(s)) = D \end{cases} \qquad [4.9]$$

where the break-even concentration λ_i is defined by: $\mu_i(\lambda_i, 0, 0) = D$.

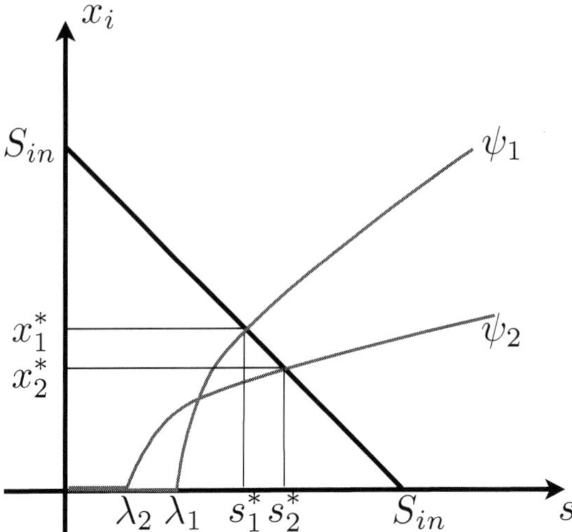

Figure 4.1. *Two characteristics at equilibrium. For a color version of this figure, see www.iste.co.uk/harmand/chemostat.zip*

If $S_{in} \leq \lambda_i$, the steady-state is the washout, otherwise there is an equilibrium with positive biomass s_i^*, x_i^* defined by:

$$x_i^* = \psi_i(s_i^*) \qquad s_i^* + x_i^* = S_{in} \tag{4.10}$$

In Figure 4.1 we have assumed that:

$$s_1^* < s_2^* \qquad \lambda_2 < \lambda_1$$

species 1 here is better "performing" than species 2 at steady-state (best biomass) but less performing for low densities (higher break-even concentration).

4.2.2. *Steady-state of two species in interaction*

For this study of model [4.6], we use the method of isoclines (see Appendix 1, section A1.4.2). Isoclines are defined by:

$$\mathcal{I}_i = \left\{ (x_1, x_2) \; : \; x_i \geq 0, \; \big(\mu_i(S_{in} - (x_1 + x_2), x_1, x_2) - D \big) x_i = 0 \right\} \tag{4.11}$$

which break down into:

$$\mathcal{I}_i = \{(x_1, x_2)\, x_j \geq 0,\ x_i = 0\ (i \neq j)\} \cup \Gamma_i \qquad [4.12]$$

with:

$$\Gamma_1 = \big\{(x_1, x_2):\ x_i\, \geq 0,\ \mu_1(S_{in} - (x_1 + x_2), x_1, x_2) - D = 0\big\}$$

$$\Gamma_2 = \big\{(x_1, x_2):\ x_i\, \geq 0,\ \mu_2(S_{in} - (x_1 + x_2), x_1, x_2) - D = 0\big\}$$

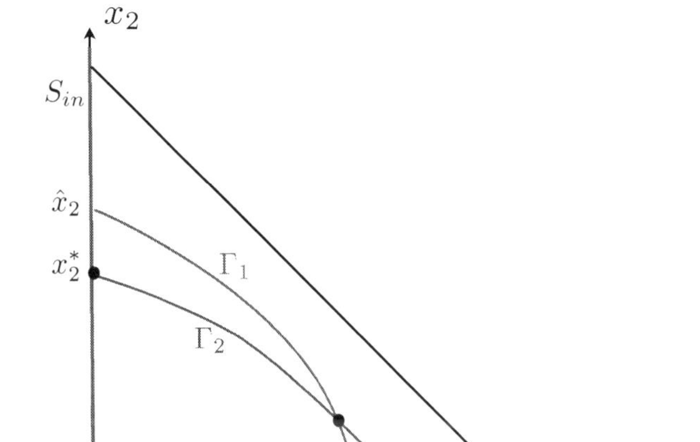

Figure 4.2. *Isoclines of [4.6]. For a color version of this figure, see www.iste.co.uk/harmand/chemostat.zip*

The intersection $\mathcal{I}_1 \cap \mathcal{I}_2$ of the two isoclines defines the steady-states.

4.2.2.1. *Washout steady-state (always)*

$(S_{in}, 0, 0)$

4.2.2.2. *Exclusion steady-states (under conditions)*

$(s_1^*, x_1^*, 0) \quad x_1^* > 0 \quad \text{if} \quad S_{in} > \lambda_1$

$(s_2^*, 0, x_2^*) \quad x_2^* > 0 \quad \text{if} \quad S_{in} > \lambda_2$

4.2.2.3. *Coexistence steady-states*

These are the potential intersection points of Γ_1 and Γ_2. To describe these two sets, we start with the following lemma 4.1:

LEMMA 4.1.–

1) if $\mu_1(S_{in}) \leq D$, the set Γ_1 is empty, otherwise it is the graph of an application $x_1 \mapsto F_1(x_1)$ defined over $[0, x_1^*[$. We denote $\hat{x}_2 = F_1(0)$; we have: $\lim_{x_1 \to x_1^*} = 0$;

2) if $\mu_2(S_{in}) \leq D$, the set Γ_2 is empty, otherwise there exists $\hat{x}_1 > 0$ and a mapping F_2 defined in $]0, \hat{x}_1]$ whose graph is Γ_2. We have $F_2(\hat{x}_1) = 0$ and $\lim_{x_1 \to 0} = x_2^*$;

3) the mappings F_1 and F_2 are strictly decreasing.

Demonstration. This is an immediate consequence of the fact that the functions $(x_1, x_2) \mapsto \mu_i(x_1, x_2)$ are strictly decreasing with respect to the two variables. We make repeated use of the intermediate value theorem. Consider the example of the set Γ_1. The mapping:

$$x_2 \mapsto \mu_1(S_{in} - x_1, \hat{x}_1, x_2)$$

is strictly decreasing because of assumptions 4.1; it decreases from the value $\mu_1(S_{in} - x_1, x_1, 0)$; the mapping:

$$x_1 \mapsto \mu_1(S_{in} - x_1, x_1, 0)$$

is strictly decreasing from value $\mu_1(S_{in})$, for $x_1 = 0$, to the value 0 for $x_1 = S_{in}$. Thus if $\mu_1(S_{in}, 0, 0) \leq D$, for all (x_1, x_2) such that: $x_1 > 0$, $x_2 > 0$, it follows that $\mu_1(S_{in} - x_1 - x_2, x_1, x_2) < D$ and Γ_1 is empty. If $\mu_1(S_{in}, 0, 0) > D$, for $x_1 < x_1^*$, it follows that $\mu_1(S_{in} - x_1, x_1, 0) > D$ and $\mu_1(S_{in} - x_1 - (S_{in} - x_1), x_1, (S_{in} - x_1)) = 0$. Consequently, there exists a unique $x_2 = F_1(x_1)$ such that $\mu_1(S_{in} - x_1 - F_1(x_1), x_1, F_1(x_1)) = D$. Thereby, $(x_1, F_1(x_1)) \in \Gamma_1$ and vice versa, which shows the point 1.

If $\mu_2(S_{in}, 0, 0) \leq D$ for all (x_1, x_2) such that: $x_1 > 0$, $x_2 > 0$ it follows that $\mu_2(S_{in} - x_1 - x_2, x_1, x_2) < D$ and Γ_2 is empty. Otherwise, there exists $\hat{x}_1$ such that for all x_1 in the interval $]0, \hat{x}_1]$, we have $\mu_2(S_{in} - x_1, x_1, 0) > D$ and since $\mu_2(S_{in} - x_1 - (S_{in} - x_1), x_1, Sin - x_1) = 0$, there exists a unique $x_2 = F_2(x_1)$ such that $\mu_2(S_{in} - x_1 - F_2(x_1), x_1, F_2(x_1)) = D$ thus $(x_1, F_2(x_1)) \in \Gamma_2$ and vice versa which demonstrates point 2.

Taking the derivative with respect to x_1 of relations $\mu_i(S_{in} - x_1 - F_1(x_1), x_1, F_i(x_1)) = D$ yields:

$$F_i' = -\frac{-\partial_s \mu_i + \partial_1 \mu_i}{-\partial_s \mu_i + \partial_2 \mu_i} \tag{4.13}$$

which is strictly negative and which proves point 2.

THEOREM 4.1.– Description of the isoclines.

1) if $\mu_2(S_{in}) \leq \mu_1(S_{in}) \leq D$, the isoclines are reduced to the two positive semi-axes. The only steady-state is washout steady-state. It is globally asymptotically stable (SAS);

2) if $\mu_2(S_{in}) \leq D < \mu_1(S_{in})$, the isoclines are reduced to the two positive semi-axes and to Γ_1 which is a curve issued from the steady-state $(x_1^*, 0)$ and that crosses the semi-axis x_2 at a point $(\hat{x}_2, 0)$ (this is the graph of F_1 defined in lemma 4.1);

3) if $D < \mu_2(S_{in}) \leq \mu_1(S_{in})$, the isoclines are constituted of the two positive semi-axes, of Γ_1 which is a curve issued from the steady-state $(x_1^*, 0)$ and that crosses the semi-axis x_2 at a point $(\hat{x}_2, 0)$ (this is the graph of F_1 defined in lemma 4.1). Furthermore, it is also constituted of Γ_2, which is a curve originating from the steady-state $(0, x_2^*)$ and that crosses the semi-axis x_1 at point $(\hat{x}_1, 0)$ (this is the graph of F_2 defined in lemma 4.1); see Figure 4.2. Curves Γ_1 and Γ_2 cannot cross each other or cross at one or several points as shown in Figure 4.3.

Demonstration. This is a paraphrase of the previous lemma.

Figure 4.3 shows all possible cases when neither Γ_1 nor Γ_2 are empty. The last two figures are examples of multiple intersections. It can be seen that assuming the simultaneous convexity (or concavity) of F_1 and F_2 does not guarantee uniqueness.

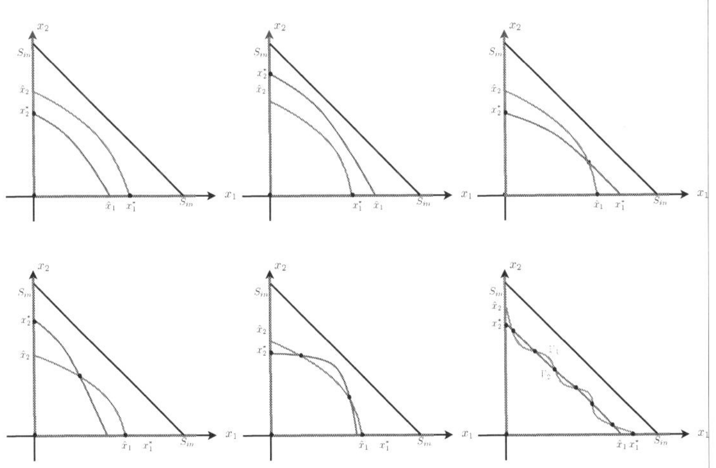

Figure 4.3. *Various possible isoclines for [4.6]. For a color version of this figure, see www.iste.co.uk/harmand/chemostat.zip*

4.2.3. *Steady-state stability*

In Figure 4.4(a), the curves Γ_1 and Γ_2 intersect one another at a single point E which is thus a coexistence steady-state; examining the arrows in the figure shows that it is GAS. Figure 4.4(b) is subtly different from the previous one: curves Γ_1 and Γ_2 have been swapped. The steady-state E is therefore always a coexistence steady-state but this time it is unstable whereas the two exclusion steady-states are stable. However, we are no longer facing with the case of competitive exclusion of Chapter 3 since here, any of the strains can win the competition. This will depend on the initial conditions.

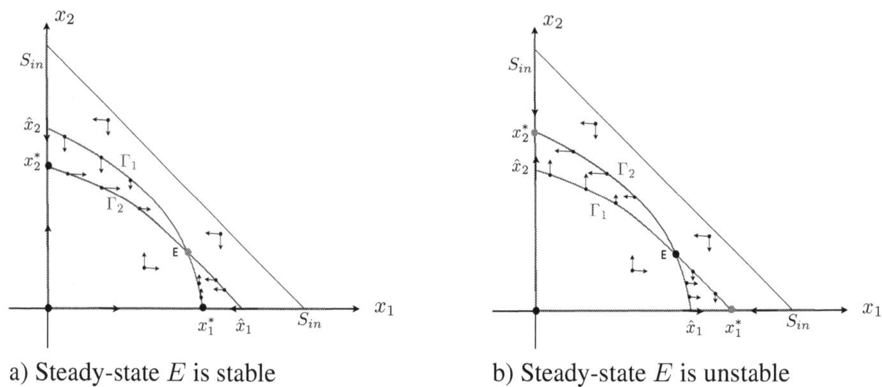

a) Steady-state E is stable   b) Steady-state E is unstable

Figure 4.4. *Coexistence steady-state stability. For a color version of this figure, see www.iste.co.uk/harmand/chemostat.zip*

Figure 4.5 represents a case where Γ_1 and Γ_2 intersect each other at four points, which thus gives four coexistence steady-states. At a steady-state point two cases are possible, the slope of the tangent to F_1 is smaller than that of the tangent to F_2 or vice versa (excluding the case where curves would be tangent). The steady-state is asymptotically stable in the first case, unstable in the second. From the formula of the derivatives of F_i (see equation [4.13]), we can conclude with the following proposition:

PROPOSITION 4.1.– A coexistence steady-state $(x_1^\dagger, x_2^\dagger)$ is asymptotically stable if and only if:

$$\frac{-\partial_s \mu_1 + \partial_1 \mu_1}{-\partial_s \mu_1 + \partial_2 \mu_1}(x_1^\dagger, x_2^\dagger) > \frac{-\partial_s \mu_2 + \partial_1 \mu_2}{-\partial_s \mu_2 + \partial_2 \mu_2}(x_1^\dagger, x_2^\dagger) \tag{4.14}$$

EXERCISE 4.1.– Find this result from the Jacobian at the steady-state point.

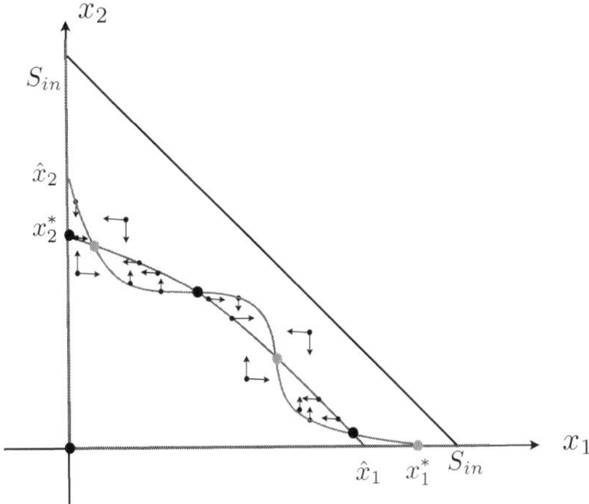

Figure 4.5. *Black steady-states are unstable;
green steady-states are stable. For a color version of this
figure, see www.iste.co.uk/harmand/chemostat.zip*

The results of this section can be partially generalized to the case where D_{x_i} are not equal to D, but then the analysis is more difficult because the reduction to dimension two can no longer be utilized (see [FEK 17]).

4.2.4. *Simulations*

Through simulations, we illustrate the results of the previous section.

The simulated model is the model:

$$\begin{cases} \dfrac{dx_1}{dt} = (\mu_1(S_{in} - (x_1 + x_2), x_1, x_2) - D)x_1 \\[2ex] \dfrac{dx_2}{dt} = (\mu_2(S_{in} - (x_1 + x_2), x_1, x_1) - D)x_2 \end{cases} \qquad [4.15]$$

with:

$$S_{in} = 2 \qquad D = 0.7$$

$$\begin{aligned} \mu_1(s, x_1, x_2) &= \frac{1.2\, s}{0.1 + s} \frac{1}{1 + (a_{11}x_1 + a_{12}x_2))^{\sigma_1}} \\[2ex] \mu_2(s, x_1, x_2) &= \frac{1.0\, s}{0.1 + s} \frac{1}{1 + (a_{21}x_1 + a_{22}x_2))^{\sigma_2}} \end{aligned} \qquad [4.16]$$

The parameters of this model are thus: a_{ij}, σ_1, and σ_2 all positive or zero. We can see that assumptions 4.1 are satisfied. The parameters a_{ii} measure the strength of the intraspecific competition, the parameters a_{ij}, $i \neq j$ measure the strength of interspecific competition.

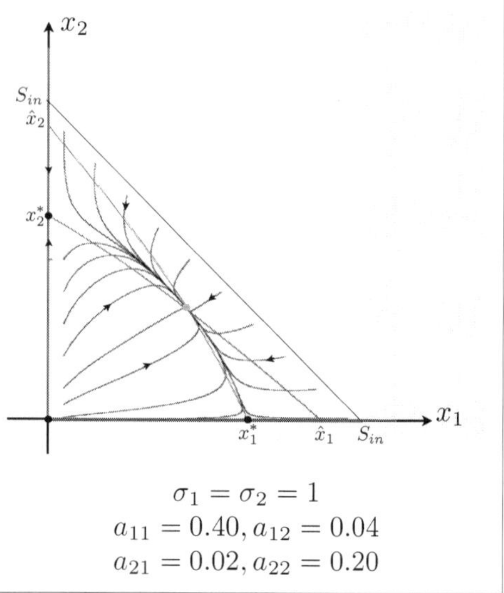

In this example Γ_1 and Γ_2 intersect at a single point where $F_1' < F_2'$; this steady-state is thus stable. The "strength" of interspecific competition (0.04 and 0.02) is much weaker than that of intraspecific competition (0.4 and 0.2).

The three steady-states: washout $((0,0))$, exclusion $((x_1^*, 0)$ and $(0, x_2^*))$ are unstable (saddles). It should be noted that the phase space is the triangle limited by the segment $x_1 + x_2 = S_{in}$.

Figure 4.6. *Coexistence steady-state. For a color version of this figure, see www.iste.co.uk/harmand/chemostat.zip*

4.3. N-species competition: exclusive intraspecific competition

Let the system be:

$$\begin{cases} \dfrac{ds}{dt} & = & D(S_{in} - s) - \sum_{i=1}^n \mu_i(s, x_i)x_i \\ \dfrac{dx_i}{dt} & = & (\mu_i(s, x_i) - D)x_i \qquad i = 1 \cdots n \end{cases} \qquad [4.17]$$

In this system, growth rates μ_i depend on s and only on the other variable x_i. We make the following assumptions:

HYPOTHESIS 4.2.–

1) for any i, we have $\mu_i(s, x_i) \geq 0$, $\mu_i(0, x_i) = 0$ and μ_i is bounded;

2) for any i, $\partial_s \mu_i > 0$, therefore the function $s \mapsto \mu_i(s, x_i)$ is strictly increasing;

3) for any i, $\partial_i \mu_i < 0$, therefore the function $s \mapsto \mu_i(s, x_i)$ is strictly decreasing.

As in the previous example, Γ_1 and Γ_2 intersect at a single point, but this time we have $F_1' > F_2'$; this steady-state is thus unstable (a saddle). The "strength" of interspecific competition (0.5 and 0.3) is much greater than those of intraspecific competition (0.1 and 0.05).

The washout steady-state $(0,0)$ is unstable (a saddle).

Exclusion steady-states $((x_1^*, 0)$ and $(0, x_2^*))$ are both stable. Their basins of attraction are separated by the two stable varieties of the saddle.

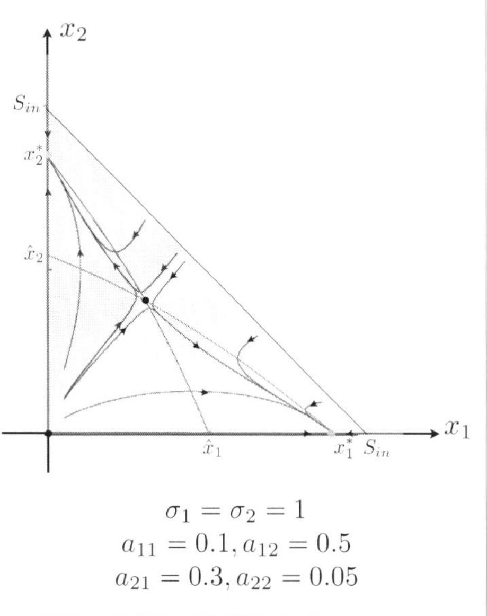

$$\sigma_1 = \sigma_2 = 1$$
$$a_{11} = 0.1, a_{12} = 0.5$$
$$a_{21} = 0.3, a_{22} = 0.05$$

Figure 4.7. *Conditional exclusion I. For a color version of this figure, see www.iste.co.uk/harmand/chemostat.zip*

In this example, we have $\sigma_2 = 10$; since $a_{22} = 1$ and $a_{21} = 0$, this results in almost stopping the growth of x_2 as soon as its concentration exceeds the value of 1. This yields an strongly curved isocline Γ_2 thereby crossing Γ_2 at two points, E_1 which is stable and E_2 which is unstable. The exclusion steady-state $(0, x_1^*)$ is also stable. Consequently, there is a possibility of coexistence associated with a possibility of exclusion.

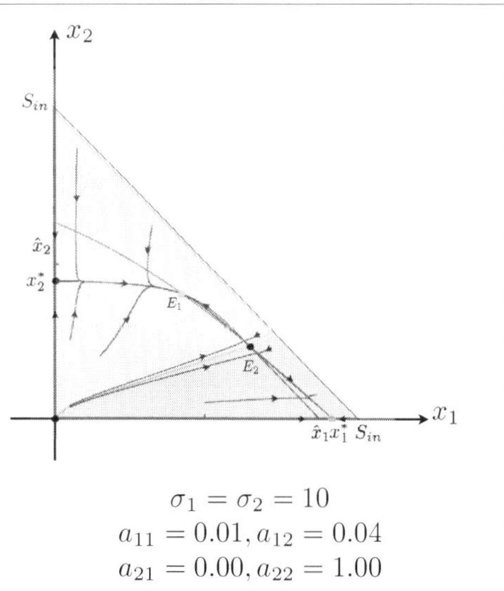

$$\sigma_1 = \sigma_2 = 10$$
$$a_{11} = 0.01, a_{12} = 0.04$$
$$a_{21} = 0.00, a_{22} = 1.00$$

Figure 4.8. *Conditional exclusion II. For a color version of this figure, see www.iste.co.uk/harmand/chemostat.zip*

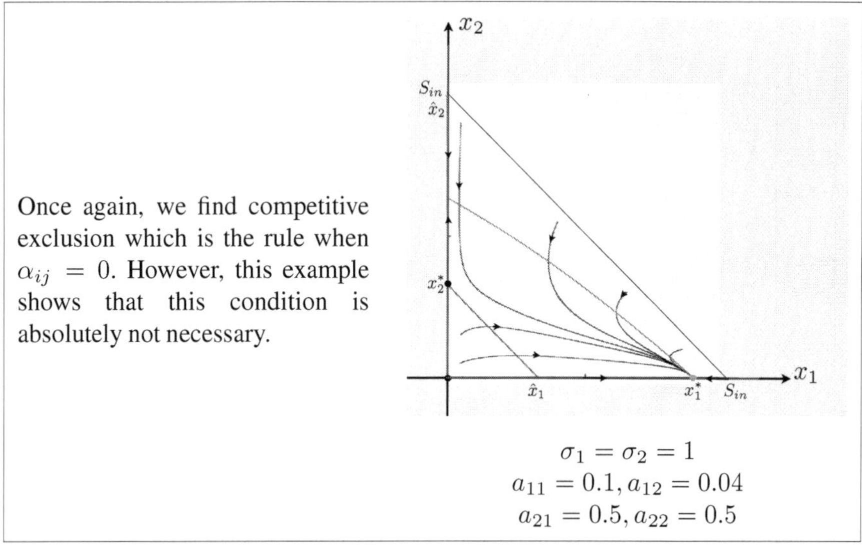

Figure 4.9. *Competitive exclusion. For a color version of this figure, see www.iste.co.uk/harmand/chemostat.zip*

The fact that each function μ_i only depends on x_i is extremely questionable but not totally unrealistic. One can easily imagine a situation where each species would be assigned to grow on a particular site (for example, beads colonized by pure species), the interference between sites being achieved only through the substrate; this would result in having intraspecific competition only. Alternatively, one might picture that the presence of biofilms organizes this kind of independence. We will see in section 4.4 how to partially move away from this hypothesis. At the moment, we consider the following as mathematics which are only relevant as such.

4.3.1. *Characteristic at equilibrium and coexistence*

In the absence of all other species, the species i is governed by the density-dependent system:

$$\begin{cases} \dfrac{ds}{dt} & = & D(S_{in} - s) - \mu_i(s, x_i)x_i \\ \dfrac{dx_i}{dt} & = & (\mu_i(s, x_i) - D)x_i \end{cases} \qquad [4.18]$$

which we have studied in detail in section 2.3.3 and that we have just mentioned again in this section 4.2.1. The characteristic of species i at equilibrium is the function:

$$s \mapsto \psi_i(s)$$

defined by:

$$\begin{aligned} s \leq \lambda_i &\implies \psi_i(s) = 0 \\ s \geq \lambda_i &\implies \mu_i(s, \psi_i(s)) = D \end{aligned} \qquad [4.19]$$

where the break-even concentration λ_i is defined by $\mu_i(\lambda_i, 0) = D$.

We know that, under the assumption that μ_i is of the "Monod" type, as soon as S_{in} is larger than the break-even concentration, there is a stable steady-state with positive biomass.

We will characterize a steady-state of [4.17] that we will call *coexistence steady-state* based on the characteristics at equilibrium of the different species. We will begin by noticing that if:

$$\left(s^*, x_i^*\right)$$

are the steady-states of each equation [4.18] corresponding to a same value s^* of s, the total consumption of the n species will be:

$$\sum_{i=1}^{n} \mu_i(s^*, x_i^*) x_i^*$$

Since we are at steady-state, it follows that $x_i^* = \psi_i(s^*)$ and therefore total consumption can be written as:

$$\sum_{i=1}^{n} \mu_i(s^*, x_i^*) x_i^* = D \sum_{i=1}^{n} \psi_i(s^*)$$

which we substitute in the equation of the dynamics of s, that is:

$$0 = D(S_{in} - s^*) - D \sum_{i=1}^{n} \psi_i(s^*)$$

thus:

$$(S_{in} - s^*) = \sum_{i=1}^{n} \psi_i(s^*)$$

On the one hand, since the function $s \mapsto S_{in} - s$ is strictly decreasing from S_{in} and the function $s \mapsto \sum_{i=1}^{n} \psi_i(s^*)$ is strictly increasing from 0, the previous equation *always has a unique solution.*

DEFINITION 4.1.– *Coexistence steady-state* of [4.17] *refers to the steady-state:*

$$(s^*, \psi_1(s^*), \psi_2(s^*) \cdot, \cdot, \cdot, \psi_i(s^*), \cdot, \cdot, \cdot, \psi_n(s^*))$$

where s^* is the unique solution of the equation:

$$(S_{in} - s^*) = \sum_{i=1}^{n} \psi_i(s^*)$$

We make some observations about this definition in Figure 4.10.

We have represented the graphs of the characteristics at equilibrium of three species ("red", "blue", "green"); the graphs of the characteristic functions are in red, green and blue. The graph of the sum of characteristics functions is in thin black. We have represented the graphs of $s \mapsto S_{in} - s$ for three values of S_{in}:

– let us start with the largest value of S_{in}, that is S_{in_1}. The intersection of the segment $[(0, S_{in_1}), \ (S_{in_1}, 0)]$ with the graph of the sum determines s_1^*; the corresponding steady-states are the intersections of the vertical line passing through s_1^* with the red, green and blue graphs, that is, in ascending order the blue, green and red points;

– for S_{in_2}, the intersection of the segment $[(0, S_{in_2}) \, , \ (S_{in_2}, 0)]$ with the graph of the sum determines s_2^* that is located to the left of the break-even concentration λ-blue of the blue species, which is therefore eliminated. Two species coexist;

– for s_{in_3}, since s_3^* is smaller than λ-green and λ-blue, the two green and blue species are eliminated at steady-state. Only the red species remains.

Since when S_{in} increases, the intersection with the graph of the sum of ψ increases indefinitely; provided that S_{in} be large enough, all species are present (that is to say, for all i, $\psi_i(s^*) > 0$). That is why this steady-state is called *coexistence steady-state.* Nonetheless, this presence of all species no longer holds when the S_{in} is too small. When S_{in} decreases, a new species is eliminated every time that s^* passes below its break-even concentration.

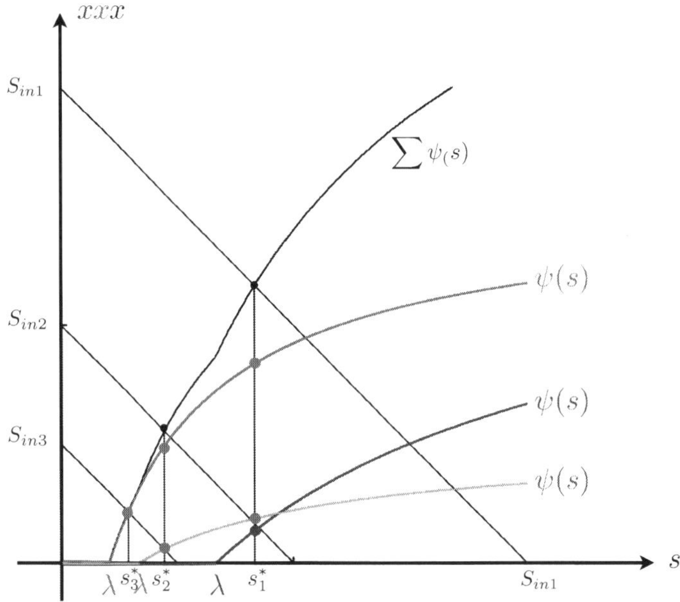

Figure 4.10. *Coexistence steady-states. For a color version of this figure, see www.iste.co.uk/harmand/chemostat.zip*

There are also other steady-states as, for example, for a given i the steady-state:

$$- \psi_i(s_i^*) = S_{in} - s^* \ ;$$
$$- x_i^* = \psi_i(s_i^*) \ ;$$
$$- x_j^* = 0 \ ; \ j \neq i \ ;$$

where all but one species are eliminated and various combinations in which certain species only are eliminated. We will not try to describe all of these possible steady-states due to the following theorem:

THEOREM 4.2.– In the presence of intraspecific competition (that is when $\forall i \ \partial_i \mu_i > 0$) and in the absence of interspecific competition, the coexistence steady-state of definition 4.17 is GAS in $(R^*)^{n+1}$.

Demonstration. The proof of this theorem goes beyond the scope of this book; it is possible to find one in [LOB 05] and another in [GRO 07].

EXERCISE 4.2.– Demonstrate that the coexistence steady-state of theorem 4.2 is locally exponentially stable.

Simulation of [4.20] + [4.21]

We have made the computer draw the three characteristics at equilibrium, their sum and the intersection point of this sum with the graph of $s \mapsto S_{in} - s$ which determines the value s^* of the substrate concentration at steady-state. The values of the concentrations of species 1, 2 and 3 at steady-state are the intersection points of the vertical line passing through s^* and the characteristics at steady-state, namely the green, blue and red points which determine the values of the concentrations at steady-state:

$$x_3^* < x_1^* < x_2^*$$

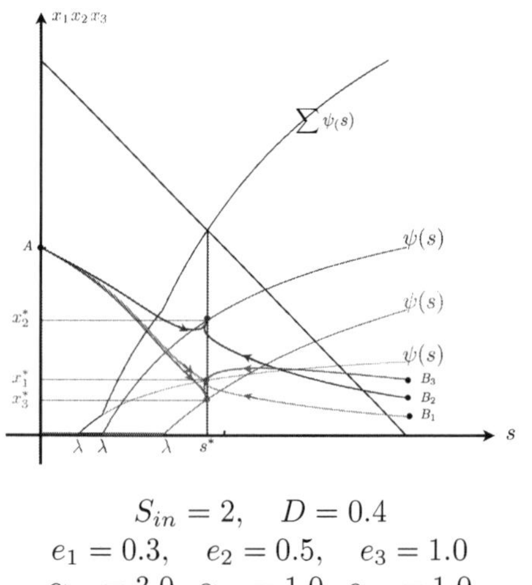

$$S_{in} = 2, \quad D = 0.4$$
$$e_1 = 0.3, \quad e_2 = 0.5, \quad e_3 = 1.0$$
$$\alpha_{11} = 3.0, \ \alpha_{22} = 1.0, \ \alpha_{33} = 1.0$$

Two trajectories have been simulated: the first originating from the point $(0, 1, 1, 1)$ (point A of the figure), the second originating from the point $(2, 0.1, 0.2, 0.3)$ (points B_1, B_2, B_3 of the figure).

Figure 4.11. *Coexistence steady-state I. For a color version of this figure, see www.iste.co.uk/harmand/chemostat.zip*

4.3.2. *Simulations*

In this section, we are going to simulate the following model:

$$\begin{cases} \dfrac{ds}{dt} & = \ D(S_{in} - s) - \sum_{i=1}^{3} \mu_i(s, x_i)x_i \\ \dfrac{dx_i}{dt} & = \ (\mu_i(s, x_i) - D)x_i \ ; \quad i = 1, 2, 3 \end{cases} \qquad [4.20]$$

with:

$$\mu_i(s, x_i) \ = \ \frac{s}{e_i + s} \frac{1}{1 + \alpha_{ii} x_i} \qquad [4.21]$$

(the notation α_{ii} replacing α_i will be justified in the next section).

The system is four-dimensional. We represent with the same axes, the three projections on the planes:

$$(os, ox_1) \quad (os, ox_2) \quad (os, ox_3)$$

A point from $\mathbb{R}^4$ therefore appears as three points *all having the same abscissa*, and consequently a trajectory of [4.20] as three curves for which we maintain the same color convention: green for x_1, blue for x_2 and red for x_3.

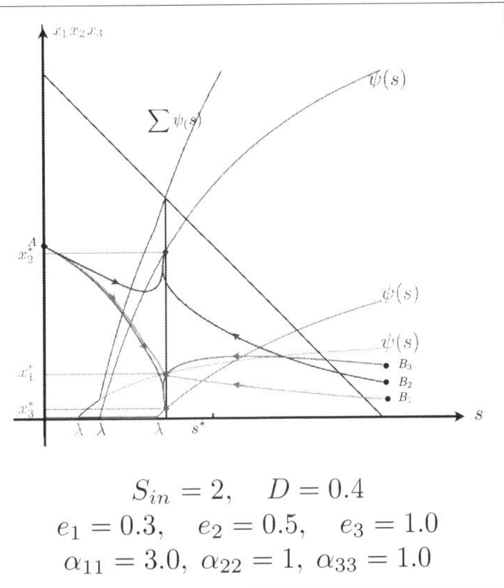

Simulation of [4.20] + [4.21]

For this simulation, all parameters are the same as those of the previous simulation except for α_{22} which is equal to 0.5 instead of 1. Therefore, we have decreased the pressure of the intraspecific competition of species 2, which has resulted in "straightening" the graph of its steady-state characteristic, thus in "straightening" the graph of the sum of the characteristic at steady-state and ultimately in decreasing s^*, x_1^* and x_3^* while x_2^* will increase.

$$S_{in} = 2, \quad D = 0.4$$
$$e_1 = 0.3, \quad e_2 = 0.5, \quad e_3 = 1.0$$
$$\alpha_{11} = 3.0, \quad \alpha_{22} = 1, \quad \alpha_{33} = 1.0$$

Figure 4.12. *Coexistence steady-state II. For a color version of this figure, see www.iste.co.uk/harmand/chemostat.zip*

4.4. N-species competition: the general case

As we have already said, the study of the general model [4.3] with any different positive $\partial_{ij}\mu_i$ and different D_{x_i} raises major problems and is not yet well understood. The mathematical study of what is known about this topic falls beyond the scope of this book. In order to simplify the study, we will consider a particular model but even for this particular model mathematical analysis remains difficult. That is the reason why we will merely comment on a few simulations.

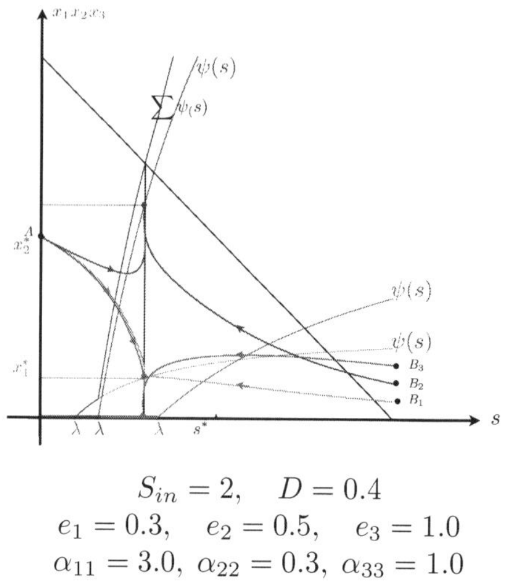

Figure 4.13. *Coexistence steady-state III. For a color version of this figure, see www.iste.co.uk/harmand/chemostat.zip*

4.4.1. *A particular density-dependent model*

We consider the model:

$$\begin{cases} \dfrac{ds}{dt} & = & D(S_{in} - s) - \sum_{i=1}^{n} \mu_i(s, x_1, \cdots, x_n) x_i \\ \dfrac{dx_i}{dt} & = & (\mu_i(s, s, x_1, \cdots, x_n) - D) x_i & i = 1 \cdots n \end{cases} \qquad [4.22]$$

with:

$$\mu_i(s, x_1, \cdots, x_n) = \nu_i(s) \frac{1}{1 + \sum_{j=1}^{n} \alpha_{ij} x_j} \qquad i = 1 \cdots n \qquad [4.23]$$

in which the functions ν_i are of the "Monod type" and $\alpha_{ij} \geq 0$. It can immediately be seen that assumptions 4.1 are satisfied. Also note that the model [4.20] of section 4.3.2 is the special case where all the α_{ij}, $i \neq j$ are zero.

The parameter α_{ii} measures the strength of the intraspecific competition within species i. The parameter α_{ij} expresses the strength of the competition pressure of species j on species i: when x_j increases, the growth rate of species i decreases. We can distinguish various types of competition:

– *exclusive-intraspecific competition*. This is the case in which all the α_{ij}, $i \neq j$ are zero; it has been studied in section 4.3; characteristics at steady-state make it possible to predict which species will be present at steady-state;

– *dominant intraspecific competition*. This is the case in which all the α_{ij}, $i \neq j$ are not zero but *small* compared to the α_{ii}: there is interspecific competition, but it is weak compared to intraspecific competition. This is a version of the previous case, which seems a bit more realistic;

– *indifferentiated competition*. This is the case in which all the α_{ij} are equal. This could be, for example, the case in which during the formation of "flocs" attachment/detachment probabilities would be independent of the species;

– *dominant interspecific competition*. In this case, the α_{ii} are small compared to the α_{ij}, $i \neq j$.

4.4.2. *Exclusive intraspecific competition*

This is the case studied in section 4.3. Review it if necessary.

4.4.3. *Dominant intraspecific competition*

It is said that a model is *robust* if a small change in its parameters does not alter its qualitative predictions. Since in our context a model is a system of differential equations, the question as to what a "robust" system differential is may arise. The appropriate mathematical concept that emerged at the end of the 1930s, is one of *structural stability* (which should not be mistaken for the stability of a steady-state). Without going into technical details, the concept is the following:

– two differential systems:

$$\frac{dx_i}{dt} = f_i(x_i) \qquad x_i \in \mathcal{D} \subset \mathbb{R}^n \tag{4.24}$$

are *equivalent* if there exists a homeomorphism ϕ (a bi-continuous bijection) of $\mathcal{D}$ such that for any trajectory $t \mapsto x_1(t)$ of the first system, the mapping $t \mapsto \phi(x_1(t))$ is a trajectory of the second. Thereby, two equivalent systems have the same number of steady-states with the same types of stability, the same number of periodic trajectories, etc.;

– a system:

$$\frac{dx}{dt} = f(x) \qquad x \in \mathcal{D} \tag{4.25}$$

is *structurally stable* if for a sufficiently small ε all systems:

$$\frac{dx}{dt} = g(x) \qquad x \in \mathcal{D} \tag{4.26}$$

such that $\|f - g\| < \varepsilon$ are equivalent to [4.25].

There are a very large number of mathematical results that help in deciding when a system of differential equations is *structurally stable*. It can be shown that this is the case of [4.22] when all the α_{ii} are strictly positive and the α_{ij}, $i \neq j$ equal to zero, and subsequently in the case of exclusive intraspecific competition. Furthermore, the introduction of "low" interspecific competition within a pure intraspecific competition model will not alter its coexistence predictions; only the numerical value of the concentrations at steady-state will be slightly modified.

Naturally, the very big drawback of the previous result is vagueness about what "low" means; in the two species case, we had a criterion but this is no longer the case for higher dimensions. In the simulations that follow, we illustrate this result by perturbing the example of Figure 4.11.

4.4.4. *Undifferentiated competition*

Contrary to the previous example, we consider the case where all the α_{ij} are equal. Assuming they are equal to 1 does not diminish generality. Therefore, we study:

$$\begin{cases} \dfrac{ds}{dt} & = & D(S_{in} - s) - \sum_{i=1}^{n} \mu_i(s, x_1, \cdots, x_n) x_i \\ \dfrac{dx_i}{dt} & = & (\mu_i(s, x_1, \cdots, x_n) - D) x_i \qquad i = 1 \cdots n \end{cases} \tag{4.27}$$

with:

$$\mu_i(s, x_1, \cdots, x_n) = \nu_i(s) \frac{1}{1 + \sum_{j=1}^{n} x_j} \qquad i = 1 \cdots n \tag{4.28}$$

Let $x = \sum_{i=1}^{n} x_i$ and consider the variables:

$$(z = s + x, \; x_1, \cdots, x_n)$$

where the system [4.27] is written as:

$$\begin{cases} \dfrac{ds}{dt} &= D(S_{in} - z) \\ \dfrac{dx_i}{dt} &= (\nu_i(z-x)\tfrac{1}{1+x} - D)x_i \qquad i = 1 \cdots n \end{cases} \qquad [4.29]$$

For a steady-state value, we should have $z = s_{in}$ and:

$$0 = \left(\nu_i(S_{in} - x)\frac{1}{1+x} - D\right)x_i \qquad i = 1 \cdots n \qquad [4.30]$$

Simulation of [4.22] + [4.23]

$$S_{in} = 2, \qquad D = 0.4$$
$$e_1 = 0.3, \qquad e_2 = 0.5, \qquad e_3 = 1.0$$

We have started by simulating the model of Figure 4.11, that is to say, the case $\alpha_{ij} = 0$, $i \neq j$.

The solution originating from the point $(0, 1, 1, 1)$ moves toward the steady-state predicted for the case of exclusive intraspecific competition; it corresponds to the three fine curves heading toward the three black points. Next, we have simulated the case with the α_{ij} indicated in the figure; they are of the order of one-tenth of α_{ii}.

The simulation issued from the same point shows curves (thicker line) that move toward a little different steady-state (the three green, blue and red points).

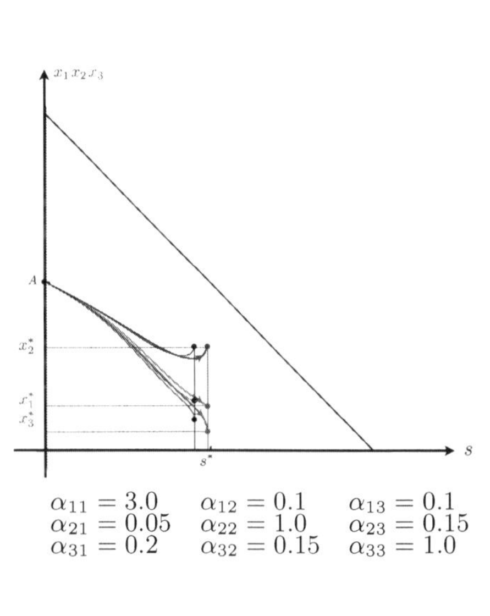

$$\begin{array}{lll} \alpha_{11} = 3.0 & \alpha_{12} = 0.1 & \alpha_{13} = 0.1 \\ \alpha_{21} = 0.05 & \alpha_{22} = 1.0 & \alpha_{23} = 0.15 \\ \alpha_{31} = 0.2 & \alpha_{32} = 0.15 & \alpha_{33} = 1.0 \end{array}$$

Figure 4.14. *Small perturbation of the model of Figure 4.11. For a color version of this figure, see www.iste.co.uk/harmand/chemostat.zip*

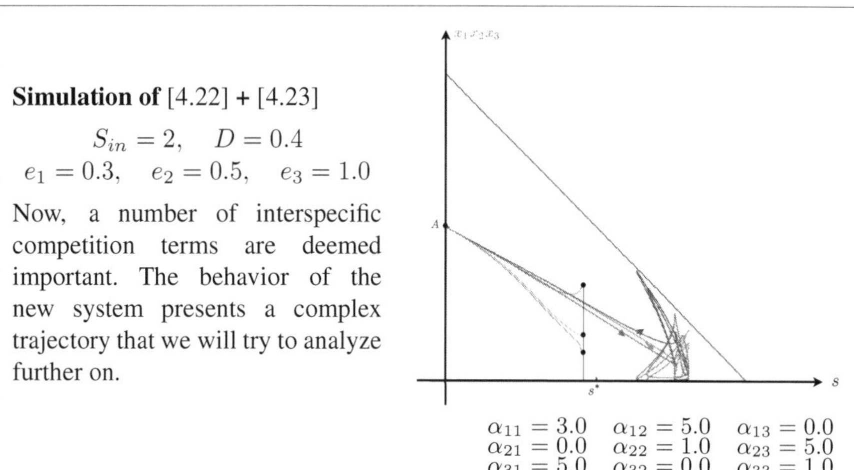

Figure 4.15. *Large perturbation of the model of Figure 4.11. For a color version of this figure, see www.iste.co.uk/harmand/chemostat.zip*

The functions $x \mapsto \nu_i(S_{in} - x)\frac{1}{1+x}$ are strictly decreasing from the value $\nu_i(S_{in})$ for $x = 0$ to the value 0 for $x = S_{in}$. The equation $\nu_i(S_{in} - x)\frac{1}{1+x} = D$ therefore has at most one solution, which allows us to define $x_i^* \; : \; i = 1 \cdots n$ by

$$\begin{cases} D \le \nu_i(S_{in}) & \Rightarrow \quad \nu_i(S_{in} - x_i^*)\frac{1}{1+x_i^*} = D \\ D \le \nu_i(S_{in}) & \Rightarrow \quad x_i^* = 0 \end{cases} \qquad [4.31]$$

PROPOSITION 4.2.– Assume that the strictly positive x_i^* are all different (which is generic) and are ordered in decreasing manner (which does not diminish the generality (see Figure 4.16):

1) the steady-states of [4.29] are washout steady-states $(S_{in}, 0, \cdots, 0)$ and steady-states:

$$(S_{in}, 0, \cdots, x_i^*, 0, \cdots, 0) \qquad [4.32]$$

2) if $D < \max \nu_i(S_{in})$, the steady-state $(S_{in}, x_1^*, 0, \cdots, 0)$ is locally exponentially stable, all other steady-states are unstable.

EXERCISE 4.3.– Demonstrate proposition 4.2.

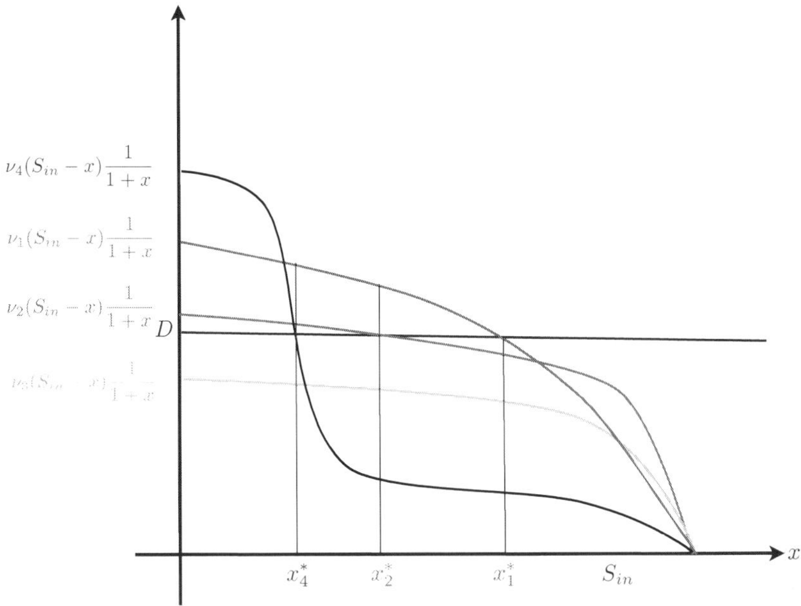

Figure 4.16. *The steady-states of [4.27]. For a color version of this figure, see www.iste.co.uk/harmand/chemostat.zip*

4.4.5. *Dominant intraspecific competition*

In Figure 4.15, we have seen a simulation of the model [4.22] + [4.23] for the values:

$$\begin{array}{lll} \alpha_{11} = 3.0 & \alpha_{12} = 5.0 & \alpha_{13} = 0.0 \\ \alpha_{21} = 0.0 & \alpha_{22} = 1.0 & \alpha_{23} = 5 \\ \alpha_{31} = 5.0 & \alpha_{32} = 0.0 & \alpha_{33} = 1.0 \end{array} \qquad [4.33]$$

for which certain interspecific competition coefficients are important. Consider the family of coefficients:

$$\begin{array}{lll} \alpha_{11} = 3.0 & \alpha_{12} = \gamma & \alpha_{13} = 0.0 \\ \alpha_{21} = 0.0 & \alpha_{22} = 1.0 & \alpha_{23} = \gamma \\ \alpha_{31} = \gamma & \alpha_{32} = 0.0 & \alpha_{33} = 1.0 \end{array} \qquad [4.34]$$

For γ ranging from 0 to 5. We thus have a family of models, indexed by γ that links an exclusive intraspecific competition model (for $\gamma = 0$) to a dominant interspecific competition model (for $\gamma = 5$). The simulations have shown that these two models behaved very differently (see Figures 4.11 and 4.15). One way to better understand the complex situation of Figure 4.15 is to vary γ in successive small increments to observe the *transition*. A value γ_0 of γ such that before and after γ_0, the systems are no longer equivalent is called a *bifurcation value*.

In our example, for $\gamma = 0$, we know that we have an exponentially stable steady-state (this is theorem 4.2); if we increase γ the eigenvalues of the Jacobian will evolve. Denote by $\lambda_1(\gamma), \lambda_2(\gamma), \lambda_3(\gamma))$ these eigenvalues and suppose that from some point in time the first two are complex conjugate:

$$\lambda_1 = a(\gamma) + ib(\gamma) \qquad \lambda_2 = a(\gamma) - ib(\gamma)$$

and finally for a value γ_0 the quantity $a(\gamma)$ which was strictly negative cancels out to become strictly positive. Then, by adding a few additional technical assumptions, it can be shown that for values greater than γ_0, there is a limit cycle whose amplitude increases with γ. This is the *Hopf bifurcation theorem* (also known as Poincaré-Andronov-Hopf theorem) which is one of the important tools for exploring differential systems. The Scholarpedia article [SCH 16] or one of the many books dedicated to the topic can be consulted, for example [MAR 12]. Software such as *Auto* [AUT 10] (which requires some training) makes it possible to numerically determine this type of bifurcation. In the following series of simulations, we can observe such bifurcation.

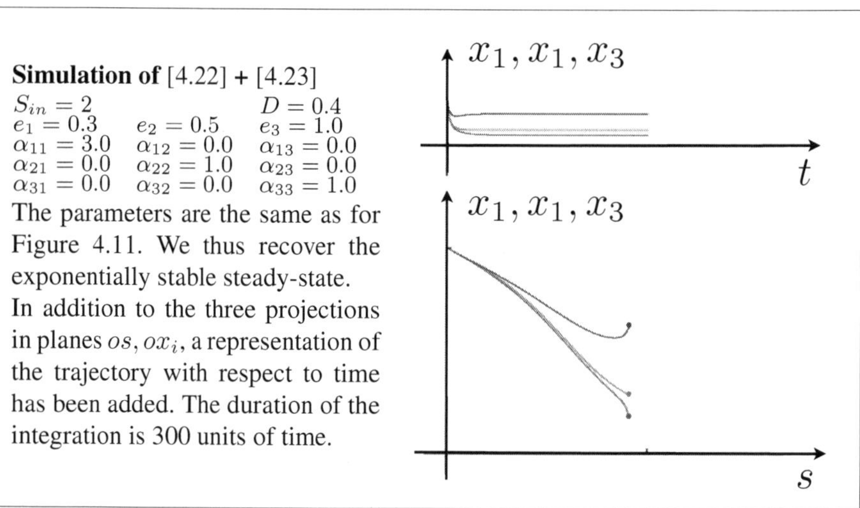

Figure 4.17. $\gamma = 0$: *steady-state. For a color version of this figure, see www.iste.co.uk/harmand/chemostat.zip*

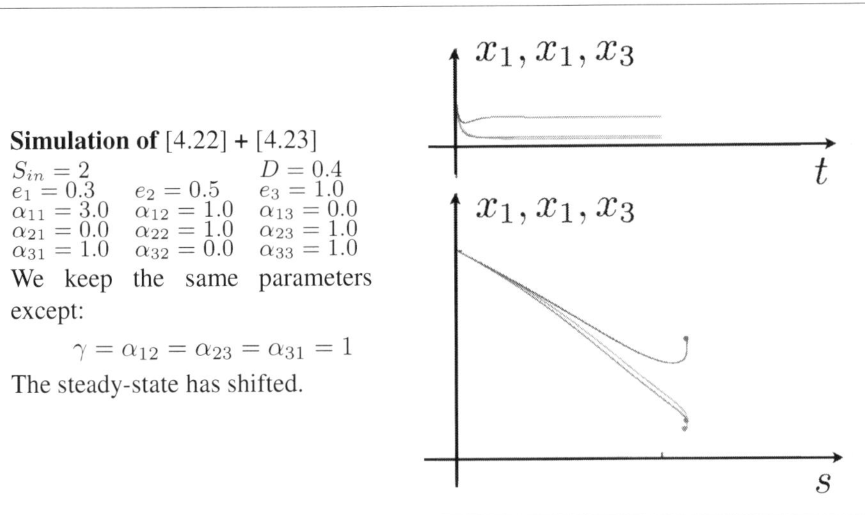

Figure 4.18. $\gamma = 1$: *steady-state. For a color version of this figure, see www.iste.co.uk/harmand/chemostat.zip*

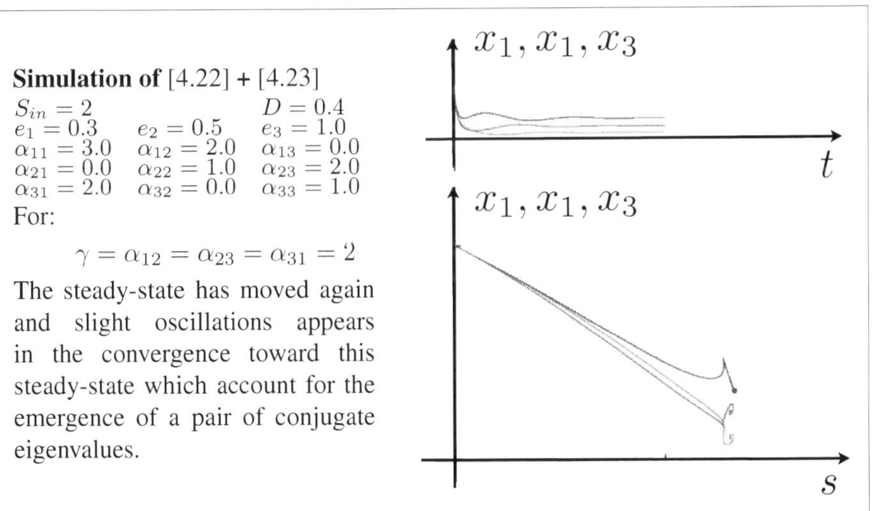

Figure 4.19. $\gamma = 2$: *beginning of the oscillations. For a color version of this figure, see www.iste.co.uk/harmand/chemostat.zip*

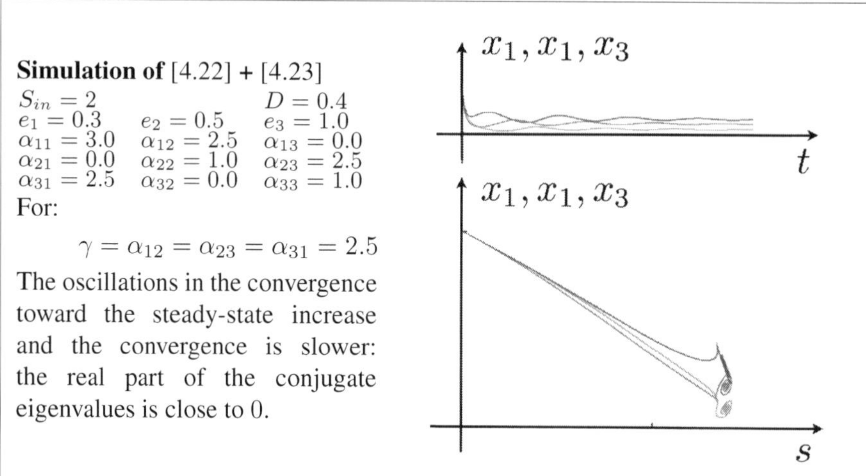

The box contains:

Simulation of [4.22] + [4.23]

$S_{in} = 2$ $D = 0.4$
$e_1 = 0.3$ $e_2 = 0.5$ $e_3 = 1.0$
$\alpha_{11} = 3.0$ $\alpha_{12} = 2.5$ $\alpha_{13} = 0.0$
$\alpha_{21} = 0.0$ $\alpha_{22} = 1.0$ $\alpha_{23} = 2.5$
$\alpha_{31} = 2.5$ $\alpha_{32} = 0.0$ $\alpha_{33} = 1.0$

For:

$$\gamma = \alpha_{12} = \alpha_{23} = \alpha_{31} = 2.5$$

The oscillations in the convergence toward the steady-state increase and the convergence is slower: the real part of the conjugate eigenvalues is close to 0.

Figure 4.20. $\gamma = 2.5$: *large oscillations. For a color version of this figure, see www.iste.co.uk/harmand/chemostat.zip*

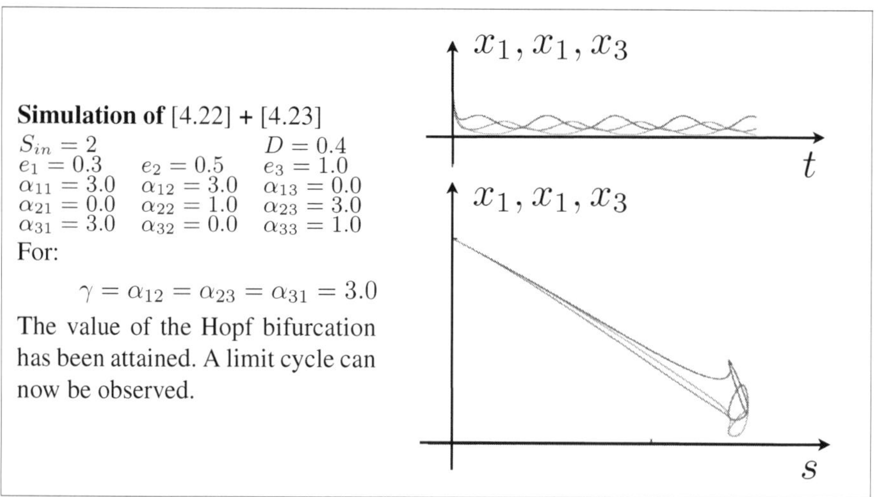

The box contains:

Simulation of [4.22] + [4.23]

$S_{in} = 2$ $D = 0.4$
$e_1 = 0.3$ $e_2 = 0.5$ $e_3 = 1.0$
$\alpha_{11} = 3.0$ $\alpha_{12} = 3.0$ $\alpha_{13} = 0.0$
$\alpha_{21} = 0.0$ $\alpha_{22} = 1.0$ $\alpha_{23} = 3.0$
$\alpha_{31} = 3.0$ $\alpha_{32} = 0.0$ $\alpha_{33} = 1.0$

For:

$$\gamma = \alpha_{12} = \alpha_{23} = \alpha_{31} = 3.0$$

The value of the Hopf bifurcation has been attained. A limit cycle can now be observed.

Figure 4.21. $\gamma = 3.0$: *limit cycle. For a color version of this figure, see www.iste.co.uk/harmand/chemostat.zip*

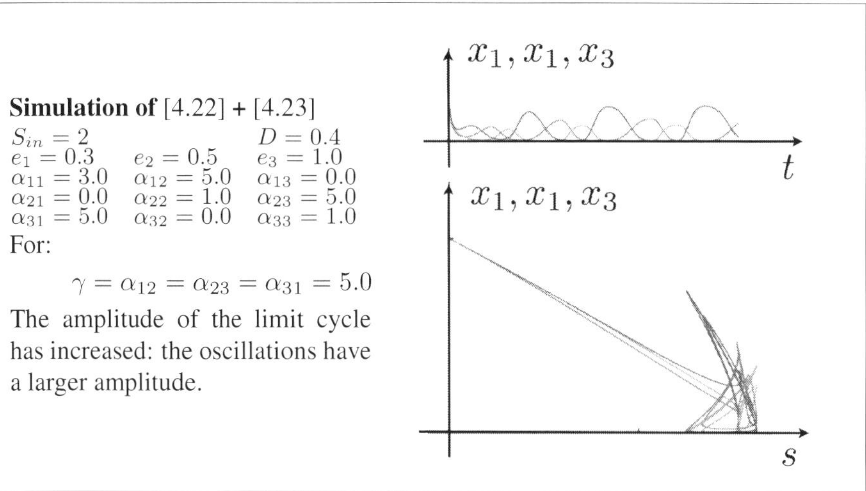

The text inside the box reads:

Simulation of [4.22] + [4.23]

$S_{in} = 2$ $D = 0.4$
$e_1 = 0.3$ $e_2 = 0.5$ $e_3 = 1.0$
$\alpha_{11} = 3.0$ $\alpha_{12} = 5.0$ $\alpha_{13} = 0.0$
$\alpha_{21} = 0.0$ $\alpha_{22} = 1.0$ $\alpha_{23} = 5.0$
$\alpha_{31} = 5.0$ $\alpha_{32} = 0.0$ $\alpha_{33} = 1.0$

For:

$$\gamma = \alpha_{12} = \alpha_{23} = \alpha_{31} = 5.0$$

The amplitude of the limit cycle has increased: the oscillations have a larger amplitude.

Figure 4.22. $\gamma = 5.0$: *limit cycle of larger amplitude. For a color version of this figure, see www.iste.co.uk/harmand/chemostat.zip*

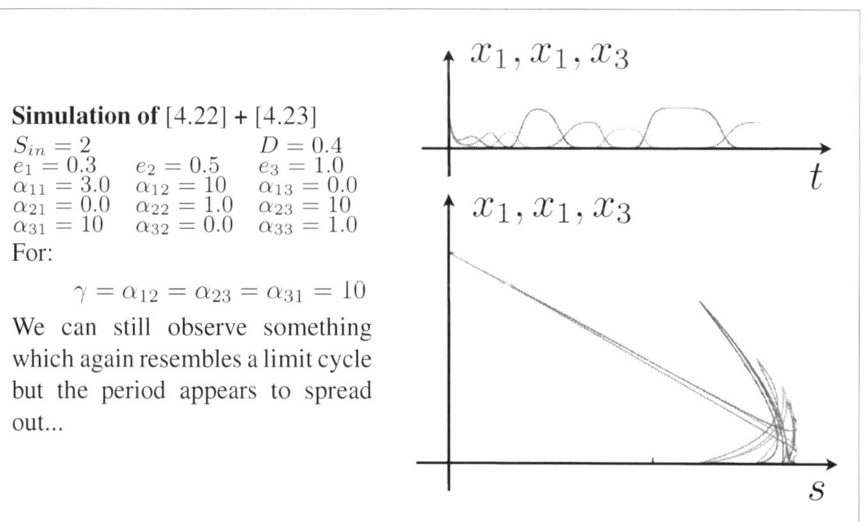

The text inside the box reads:

Simulation of [4.22] + [4.23]

$S_{in} = 2$ $D = 0.4$
$e_1 = 0.3$ $e_2 = 0.5$ $e_3 = 1.0$
$\alpha_{11} = 3.0$ $\alpha_{12} = 10$ $\alpha_{13} = 0.0$
$\alpha_{21} = 0.0$ $\alpha_{22} = 1.0$ $\alpha_{23} = 10$
$\alpha_{31} = 10$ $\alpha_{32} = 0.0$ $\alpha_{33} = 1.0$

For:

$$\gamma = \alpha_{12} = \alpha_{23} = \alpha_{31} = 10$$

We can still observe something which again resembles a limit cycle but the period appears to spread out...

Figure 4.23. $\gamma = 10$: *limit cycle. For a color version of this figure, see www.iste.co.uk/harmand/chemostat.zip*

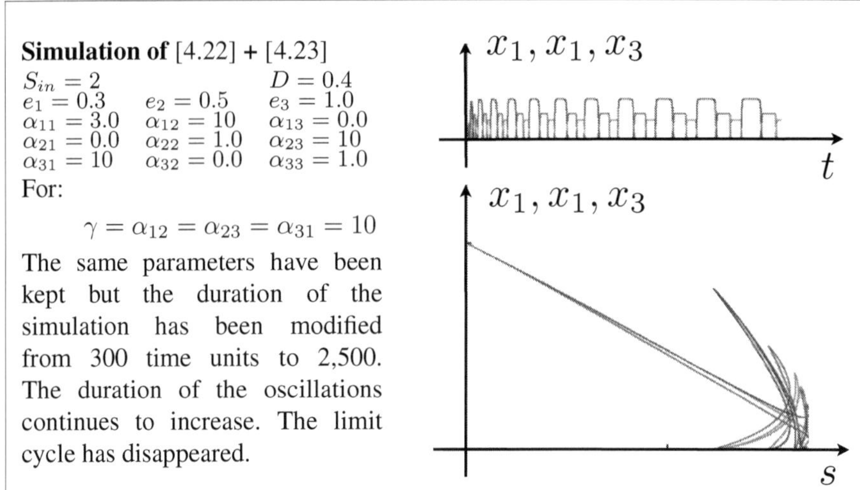

Simulation of [4.22] + [4.23]

$S_{in} = 2$ $D = 0.4$
$e_1 = 0.3$ $e_2 = 0.5$ $e_3 = 1.0$
$\alpha_{11} = 3.0$ $\alpha_{12} = 10$ $\alpha_{13} = 0.0$
$\alpha_{21} = 0.0$ $\alpha_{22} = 1.0$ $\alpha_{23} = 10$
$\alpha_{31} = 10$ $\alpha_{32} = 0.0$ $\alpha_{33} = 1.0$

For:

$$\gamma = \alpha_{12} = \alpha_{23} = \alpha_{31} = 10$$

The same parameters have been kept but the duration of the simulation has been modified from 300 time units to 2,500. The duration of the oscillations continues to increase. The limit cycle has disappeared.

Figure 4.24. $\gamma = 10$: simulation duration 2,500. For a color version of this figure, see www.iste.co.uk/harmand/chemostat.zip

This latest simulation, for $\gamma = 10$, has an interesting interpretation. Consider the two species 1 and 2 in the absence of species 3. Its model is:

$$\begin{cases} \dfrac{ds}{dt} & = & D(S_{in} - s) - \dfrac{s}{0.3 + s}\dfrac{1}{1 + 3\,x_1 + 10\,x_2}x_1 - \dfrac{s}{0.5 + s}\dfrac{1}{1 + 1\,x_2}x_2 \\[2ex] \dfrac{dx_1}{dt} & = & \left(\dfrac{s}{0.3 + s}\dfrac{1}{1 + 3\,x_1 + 10\,x_2} - D \right)x_1 \\[2ex] \dfrac{dx_2}{dt} & = & \left(\dfrac{s}{0.5 + s}\dfrac{1}{1 + x_2} - D \right)x_2 \end{cases} \qquad [4.35]$$

It can easily be verified that for this model species 2 eliminates species 1. In fact, species 2 experiences intraspecific competition only while it exerts a very strong pressure ($\gamma = 10$) on species 1. It is as easily verified that when species 2 is not present, it is species 1 that eliminates species 3 and finally in the absence of species 1 that it is now 3 which eliminates no. 2. Thereby, we have:

1 eliminates 3 which eliminates 2 which eliminates 1

Let us start with a high concentration of 1 and 3 as well as a low concentration of 2. At the beginning, everything goes as if 1 and 3 were alone, therefore 1 tends toward its steady-state value and 3 tends toward 0. On the other hand, 3 does no longer exert pressure on 2 which can grow and begins to eliminate 1, which will allow 3 to grow, etc. This is what is observed in Figure 4.24 where each species alternately constitutes, and for increasingly longer periods, most of the biomass. Of course, in reality, such a mechanism cannot last forever because the concentrations of each species take increasingly smaller values that make the model quickly totally inadequate.

4.5. Bibliographic notes

The mathematical article [DEL 06], whose title *Crowding effects promote coexistence in the chemostat* is explicit, has highlighted the effect of intraspecific competition induced by high densities. For a complete and detailed study of this model see [ABD 16].

Competition models in a chemostat with μ_i being density-dependent have been considered particularly by [ANG 04, DEL 06, FEK 13b, GRO 07, LOB 05, LOB 06a, LOB 06b]. In fact, methods originating from systems theory have been introduced by [ANG 04, DEL 06] to address these issues. In addressing this chapter, we have essentially followed [LOB 06a, LOB 06b] and [FEK 13b]. The "asymptotic" model:

$$\begin{cases} \dfrac{dx_i}{dt} & = & (\mu_i(S_{in} - \sum_{j=1}^{n} x_j, x) - D)x_i \\ x = (x_1, \cdots x_n) \quad i = 1 \cdots n \end{cases} \qquad [4.36]$$

is of the form:

$$\begin{cases} \dfrac{dx_i}{dt} & = & f_i(x_1, \cdots, x_n)x_i \qquad i = 1 \cdots n \end{cases} \qquad [4.37]$$

and under our assumptions, it follows that:

$$\frac{\partial f_i}{\partial x_j} = -\mu_j' + \partial_{ij}\mu_i \leq 0 \qquad [4.38]$$

Systems of this type are called *competitive systems* and abundant mathematical literature (initiated by M. Hirsch [HIR 82, HIR 85, HIR 88]) is dedicated to them. Despite the very strong condition [4.38], competitive systems can have very different asymptotic behaviors as Smale demonstrated in 1976 in a short article [SMA 76]. In particular, May and Leonard have shown in 1975 [MAY 75] how a Lotka-Volterra system with three competitors could present a periodic solution.

5

More Complex Models

5.1. Introduction

In this final chapter, we will try to provide insight about situations more complex than the ones we have proposed in Chapters 2, 3 and 4. Indeed, there are countless models that complexify the basic model in several directions:

– more accurately taking into account the biology of bacteria: flocculation, biofilm constitution, etc.;

– presence of bacteria "consumers": viruses, amoebae, etc.;

– presence of multiple resources;

– spatially distributed resources and populations;

– physiologically structured individuals: age, height, etc.;

– variable environment;

– explicitly taking births and deaths into account (stochastic models);

– taking evolution into account, etc.

Several books would be required to account for all these extensions. Here, we will only review two types of model that attempt to model the first two situations above. The mathematical treatment is therefore more succinct and we refer the reader to the literature for more details.

5.2. Models with aggregated biomass

Under certain growth conditions and in some environments, microbial species may present aggregates of microorganisms or *flocs* of various sizes (see Figure 5.1). Microorganisms can also attach themselves to the walls of tanks, pipes, reactors, etc. (or more generally of any chemostat-based device), and thus create *biofilms* with varied thicknesses. Over time, micro-organisms, parts of flocs or of biofilms detach and dump in the liquid medium isolated individuals or small-sized aggregates (see Figure 5.2). These bacterial assemblages (which can be observed under the microscope) affect the performance of chemostats at the macroscopic level, namely regarding:

– the growth of biomass: bacterial individuals have differentiated access to substrate depending on their position inside or on the periphery of assemblies. In addition, microorganism secretions of polymers that enable the attachment are generally achieved to the detriment of their growth;

– the disappearance of biomass: flocs and biofilms are most often less likely to be dragged away by the chemostat outflow, comparatively to isolated individuals.

The appearance and evolution mechanisms of these assemblies, which at the same time relate to biology, mechanics and hydrodynamics, are complex, partially understood and difficult to model at a microscopic scale. The objective of this chapter is to study how the conventional model of the chemostat can be enriched with considerations reflecting the effects of biomass attachment and detachement at the macroscopic level (in other words, without representing all the refinements that a description would bring at the microscopic level).

Figure 5.1. *Isolated individuals may aggregate to form a floc, or else attach to an already formed aggregate. For a color version of this figure, see www.iste.co.uk/harmand/chemostat.zip*

Figure 5.2. *Individuals can detach from an aggregate. An aggregate can be split into smaller aggregates. For a color version of this figure, see www.iste.co.uk/harmand/chemostat.zip*

5.2.1. *Planktonic biomass* versus *aggregate biomass*

We consider that the total biomass of a given species is decomposed into "planktonic" (or "free") biomass made up of non-attached microorganisms (or at least those that behave as such; which may still be the case of small assemblies) and "aggregate" biomass (without accurately taking account of the shape and of the size of assemblies). Thus, we write the concentration x of the total biomass as the sum of concentrations u and v of planktonic and aggregate biomass, respectively:

$$x = u + v \qquad [5.1]$$

This distinction allows us to take into account different growth and death characteristics according to whether microorganisms are attached or not. We thus denote respectively by $\mu_u(\cdot)$, D_u and $\mu_v(\cdot)$, D_v the specific growth and removal rates of planktonic and aggregate compartments. On the other hand, we denote the specific velocity of attachment of planktonic biomass by $\alpha(\cdot)$ and that of detachment of the attached biomass by $\beta(\cdot)$. As a result, we obtain the following model:

$$\begin{cases} \frac{ds}{dt} = D(S_{in} - s) - \mu_u(s)u - \mu_v(s)v \\[2mm] \frac{du}{dt} = \mu_u(s)u - D_u u - \alpha(\cdot)u + \beta(\cdot)v \\[2mm] \frac{dv}{dt} = \mu_v(s)v - D_v v + \alpha(\cdot)u - \beta(\cdot)v \end{cases} \qquad [5.2]$$

As previously, we take unit yield coefficients without loss of generality. The simplicity of this representation, which does not account for the richness of forms and possible sizes of aggregates, should be regarded as the consideration of an *average* microorganism behavior within aggregates or biofilms, which differs from that of isolated microorganisms. Since it is difficult to obtain or to justify precise expressions of the attachment and detachment terms for this type of model, our purpose is to understand and qualitatively predict the possible effects of these terms on the dynamics of the system (to this end, we will merely consider simple expressions as possible representatives). It should be noted that the attachment and detachment terms depend on the operating conditions (in particular the dilution rate). This is why in this chapter we consider fixed operating conditions.

Hereafter, we make the following assumptions, which reflect the considerations discussed in the introduction:

HYPOTHESIS 5.1.–

1) The specific growth kinetics $\mu_u(\cdot)$ and $\mu_v(\cdot)$ are concave increasing functions that verify:

$$\mu_u(s) > \mu_v(s), \quad \forall s > 0 \tag{5.3}$$

2) The removal rates of aggregate and planktonic biomass verify:

$$D_u \geq D_v \tag{5.4}$$

3) The function α only depends on concentrations u and v in an increasing manner and such that:

$$u > 0 \implies \alpha(u, 0) > 0$$

with:

$$\frac{\partial \alpha}{\partial u}(u, v) \geq \frac{\partial \alpha}{\partial v}(u, v), \quad \forall(u, v)$$

4) The function β depends only on the concentration v in a decreasing manner and such that $v \mapsto \beta(v)v$ is increasing with:

$$v > 0 \implies \beta(v) > 0$$

Typically, it can be considered that the specific attachment velocity $\alpha(u, v)$ can be decomposed into a sum of two terms $\alpha_u(u)$ and $\alpha_v(v)$ that reflect the two possible types of attachments: on free bacteria or on bacteria already in flocs. Considering that free bacteria mainly attach on the surface of flocs, and that when the size of flocs increases, the ratio surface over volume does not increase as quickly as the volume, it can be expected that the function α_v increases more slowly than α_u, which is then reflected by $\alpha'_u(u) \geq \alpha'_v(v)$ for all (u, v), justifying assumption 3).

In general, it is expected that the detachment velocity $v \mapsto \beta(v)v$ increases with the density v of the attached biomass, but when the floc size increases, the ratio surface over volume increases more slowly than the volume, which results in a decrease of the function β, thus justifying assumption 4.

5.2.2. *Coexistence between the two forms*

We will assume that:

$$D = D_u = D_v$$

which allows us, as in Chapter 3, to consider the variable $z(t) = s(t) + x(t)$, solution of the differential equation:

$$\frac{dz}{dt} = D(S_{in} - z)$$

Thus, after a transient, the solution trajectories of the system [5.2] are approximated by the system reduced to two dimensions:

$$\begin{cases} \frac{du}{dt} = \mu_u(S_{in} - u - v)u - Du - \alpha(u,v)u + \beta(v)v \\ \frac{dv}{dt} = \mu_v(S_{in} - u - v)v - Dv + \alpha(u,v)u - \beta(v)v \end{cases} \qquad [5.5]$$

We study the possible steady-states $(u^\star, v^\star)$ of this system, that is to say, the positive solutions of the system:

$$\begin{cases} \mu_u(S_{in} - u - v)u - Du - \alpha(u,v)u + \beta(v)v = 0 \\ \mu_v(S_{in} - u - v)v - Dv + \alpha(u,v)u - \beta(v)v = 0 \end{cases} \qquad [5.6]$$

It can be immediately noticed that $u^\star = 0$ implies $\beta(v^\star)v^\star = 0$ and $v^\star = 0$ implies $\alpha(u^\star, 0)u^\star = 0$. The assumptions 5.1 that we consider on terms $\alpha(\cdot)$ and $\beta(\cdot)$ then allow us to infer that there is no steady-state where only one of the two forms would be present.

5.2.3. *Coexistence steady-state*

By adding up these two equations and taking [5.1] into account, we obtain $(u^\star, v^\star)$ as a solution of the system:

$$\begin{cases} (\mu_u(s) - D)u + (\mu_v(s) - D)v = 0 \\ u + v = S_{in} - s \end{cases}$$

For example, a coexistence steady-state (if it exists) verifies:

$$u^\star = (S_{in} - s^\star)\frac{D - \mu_v(s^\star)}{\mu_u(s^\star) - \mu_v(s^\star)}, \quad v^\star = (S_{in} - s^\star)\frac{\mu_u(s^\star) - D}{\mu_u(s^\star) - \mu_v(s^\star)} \qquad [5.7]$$

According to hypothesis [5.3], we obtain the following necessary condition:

$$\mu_u(s^\star) > D > \mu_v(s^\star)$$

By defining the break-even concentration by λ_u, λ_v as the dilution rate D, as we did in Chapter 3 (that is that verify $\mu_u(\lambda_u) = \mu_v(\lambda_v) = D$ with $\lambda_v > \lambda_u$), we deduce that a coexistence steady-state must verify:

$$s^\star \in [\lambda_u, \lambda_v]$$

Thus, a necessary condition for the existence of a coexistence steady-state is:

$$\lambda_u < S_{in} \tag{5.8}$$

At this stage, it is difficult to prove the existence of solutions without specifying attachment and detachment functions $\alpha(\cdot)$ and $\beta(\cdot)$. If we consider that we are only dealing with flocs of small sizes, as a first approximation it is possible to assume that α is a function of $x = u + v$ (that is, functions α_u and α_v are identical), which will be chosen as linear (to simplify), and that the function β does not depend on v:

$$\alpha(u, v) = a(u + v) = ax, \quad \beta(v) = b \tag{5.9}$$

where a and b are two positive constants. Thereby, hypotheses 5.1 are verified.

PROPOSITION 5.1.– For growth functions μ_u, μ_v that verify point 1) of assumption 5.1 and attachment and detachment functions $\alpha(\cdot)$, $\beta(\cdot)$ of the form [5.9], there exists a unique coexistence steady-state if and only if the condition:

$$D < \mu_u(S_{in}) \tag{5.10}$$

is verified.

Demonstration. I denotes the interval:

$$I =]\lambda_u, \lambda_v[.$$

To simplify the writing, the following notations are introduced:

$$\varphi_u(s) = \mu_u(s) - D \quad \text{et} \quad \varphi_v(s) = \mu_v(s) - D$$

For all $s \in I$, we have $\varphi_u(s) > 0 > \varphi_v(s)$. The steady-states (s^*, u^*, v^*) are given by:

$$\begin{cases} 0 = \varphi_u(s^*)u^* - a(u^* + v^*)u^* + bv^* \\ 0 = \varphi_v(s^*)v^* + a(u^* + v^*)u^* - bv^* \end{cases} \tag{5.11}$$

with $u^* + v^* = S_{in} - s^*$. If $u^* = 0$ then, from the first equation, it can be deduced that $v^* = 0$. Similarly, if $v^* = 0$ then, from the second equation it can be deduced that $u^* = 0$. Consequently, the steady-states are the washout $E_0 = (S_{in}, 0, 0)$ and a steady-state of the form:

$$E^* = (s^*, u^*, v^*)$$

with $u^* > 0$, $v^* > 0$ and $s^* = S_{in} - u^* - v^*$.

In order to solve equations [5.11], one uses a similar method to the method of the characteristic at steady-state used in Chapters 2 and 4. We recall that this method consists of determining the steady-states of the system formed by the 2nd and 3rd equations of [5.2], where the variable s is considered to be an input of the system. In other words the aim is to solve the system formed by the first and the second equation of [5.11], in which u^* and v^* are the unknowns and s^* is considered as being a parameter. It thus yields:

$$u^* = U(s^*), \qquad v^* = V(s^*)$$

If u^* and v^* are replaced by these expressions in the first equation of [5.2], an equation of the single variable s^* is obtained of the form:

$$D(S_{in} - s^*) = H(s^*) \quad \text{with} \quad H(s^*) = \mu_u(s^*)U(s^*) + \mu_v(s^*)V(s^*)$$

that is solved, see Figure 5.3, to find a positive solution s^*. This solution gives a positive steady-state, provided that $U(s^*)$ and $V(s^*)$ are positive. In the following, the functions U, V and H are determined and the conditions are given in order for the solution s^* to exist.

By summing the 1st and 2nd equations [5.11], we obtain:

$$\varphi_u(s^*)u^* + \varphi_v(s^*)v^* = 0 \tag{5.12}$$

This equation admits a positive solution if and only if $\varphi_u(s^*)$ and $\varphi_v(s^*)$ are of opposite signs, that is, if and only if $s^* \in I$. If this equation admits a solution in this interval then equation [5.12] can be written as follows:

$$v^* = -\frac{\varphi_u(s^*)}{\varphi_v(s^*)}u^* \tag{5.13}$$

By replacing v^* by Expression [5.13] in the first equation of [5.11], it yields:

$$u^* = U(s^*) \quad \text{with} \quad U(s) = \frac{\varphi_u(s)(\varphi_v(s) - b)}{a[\varphi_v(s) - \varphi_u(s)]} \tag{5.14}$$

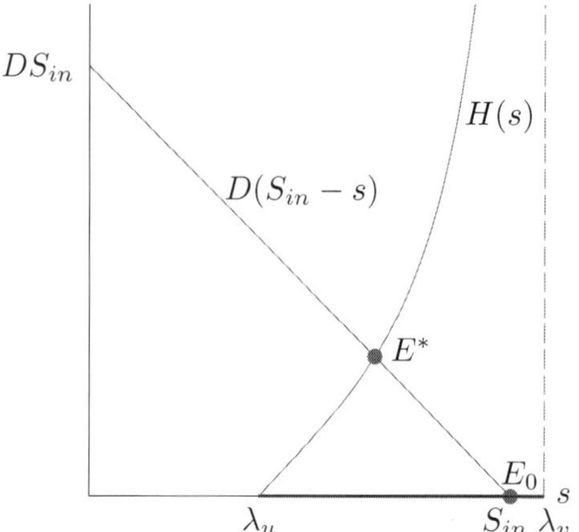

Figure 5.3. *Existence of a unique positive steady-state. For a color version of this figure, see www.iste.co.uk/harmand/chemostat.zip*

Note that u^* defined by [5.14] is positive because $s^* \in I$. By replacing u^* by [5.14] in [5.13], we get:

$$v^* = V(s^*) \quad \text{with} \quad V(s) = -\frac{\varphi_u^2(s)(\varphi_v(s) - b)}{a[\varphi_v(s) - \varphi_u(s)]\varphi_v(s)} \tag{5.15}$$

Substituting the expressions of $U(s^*)$ and $V(s^*)$ given by by [5.14] and [5.15] in the expression of $H(s^*)$ yields a characterization of s^*:

$$D(S_{in} - s^*) = H(s^*) \quad \text{with} \quad H(s) = D\frac{\varphi_u(s)(\varphi_v(s) - b)}{a\varphi_v(s)} \tag{5.16}$$

Note that for all $s \in I$, we have $U(s) > 0$, $V(s) > 0$ and $H(s) > 0$, and that:

$$\lim_{s \to \lambda_u} H(s) = 0, \quad \lim_{s \to \lambda_v} H(s) = +\infty$$

In addition, function H is strictly increasing on I. Indeed, we have:

$$H'(s) = \frac{D}{a}\frac{\varphi_v(s)(\varphi_v(s) - b)\varphi_u'(s) + b\varphi_u(s)\varphi_v'(s)}{\varphi_v^2(s)} > 0$$

Consequently, equation [5.16] admits a unique solution $s^* \in I =]\lambda_u, \lambda_v[$ if and only if $S_{in} > \lambda_u$, which is equivalent to $\mu_u(S_{in}) > D$.

5.2.4. *Stability study*

Under the conditions of stability and global attractiveness of the washout steady-state of the chemostat model in which only the planktonic biomass would be considered:

$$D \geq \mu_u(S_{in}) \qquad [5.17]$$

(see Chapter 2), the washout is also the only steady-state of the system [5.2], stable and globally attractive. As a matter of fact, by considering the reduced model [5.5], under this assumption we have:

$$x \in\,]0, S_{in}] \;\Rightarrow\; \frac{dx}{dt} = (\mu_u(S_{in} - x) - D)u + (\mu_v(S_{in} - x) - D)v < 0$$

which demonstrates that $x(\cdot)$ asymptotically converges toward 0 for any initial condition.

According to the study conducted in section 5.2.3, a positive steady-state exists as soon as the condition [5.10] is verified and is unique. By particularizing the attachment and detachment functions as we did in section 5.2.3, the following stability result is obtained (the case in which D_u and D_v are different from D is addressed in [FEK 13a]).

PROPOSITION 5.2.– Under the assumptions of proposition 5.1. the coexistence steady-state is locally exponentially stable.

Demonstration. The Jacobian matrix of [5.5] for the steady-state (u^*, v^*), which corresponds to the positive equilibrium $E^* = (s^*, u^*, v^*)$ of [5.2], is equal to:

$$J^* = \begin{bmatrix} -u^* \varphi_u'(s^*) + \varphi_u(s^*) - a(2u^* + v^*) & -u^* \varphi_u'(s^*) - au^* + b \\ -v^* \varphi_v'(s^*) + a(2u^* + v^*) & -v^* \varphi_v'(s^*) + \varphi_v(s^*) + au^* - b \end{bmatrix}$$

The trace of this matrix is equal to:

$$\mathrm{Tr}\, J^* = -u^* \varphi_u'(s^*) - v^* \varphi_v'(s^*) + \varphi_u(s^*) - a(u^* + v^*) + \varphi_v(s^*) - b$$

Note that based on equations [5.11], it can be deduced that:

$$\varphi_u(s^*) - a(u^* + v^*) = -b\frac{v^*}{u^*} < 0, \qquad \varphi_v(s^*) - b = -a\frac{(u^* + v^*)u^*}{v^*} < 0 \quad [5.18]$$

Further, as $\varphi_u'(s^*) > 0$ and $\varphi_v'(s^*) > 0$, it can be deduced that $\mathrm{Tr}\, J^* < 0$.

The determinant of this matrix is equal to:

$$\mathrm{Det}\,J^* = Au^*\varphi_u'(s^*) + Bv^*\varphi_v'(s^*) + C$$

with:

$$A = a(u^* + v^*) + b - \varphi_v(s^*), \quad B = a(u^* + v^*) + b - \varphi_u(s^*)$$

and:

$$C = \varphi_u(s^*)\varphi_v(s^*) + \varphi_u(s^*)(au^* - b) - \varphi_v(s^*)a(2u^* + v^*)$$

By using expressions [5.18], it yields that:

$$A = a\frac{(u^* + v^*)^2}{v^*} > 0, \quad B = b\frac{u^* + v^*}{u^*} > 0$$

Moreover, we have:

$$C = \varphi_u(s^*)\left(\varphi_v(s^*) - b\right) + a\left(u^*\varphi_u(s^*) - v^*\varphi_v(s^*)\right) - 2au^*\varphi_v(s^*)$$

Using [5.12], we get:

$$C = \varphi_u(s^*)\left(\varphi_v(s^*) - b\right) + 2au^*\varphi_u(s^*) - 2au^*\varphi_v(s^*)$$

Utilizing [5.14], we have:

$$au^*\left(\varphi_u(s^*) - \varphi_v(s^*)\right) = -\varphi_u(s^*)\left(\varphi_v(s^*) - b\right)$$

Consequently:

$$C = -\varphi_u(s^*)\left(\varphi_v(s^*) - b\right) > 0$$

Therefore, it can be deduced that $\mathrm{Det}\,J^* > 0$, and as a consequence, the real parts of the eigenvalues are strictly negative.

5.2.5. *The case of fast attachments/detachments*

Depending on species and on hydrodynamic conditions, attachment and detachment velocities may prove to be large compared to growth kinetics and to the dilution rate. In this case, it is possible to consider that the attachment and detachment terms, $\alpha(\cdot)$ and $\beta(\cdot)$ respectively, can be rewritten in the form:

$$\frac{\alpha(\cdot)}{\varepsilon}, \quad \frac{\beta(\cdot)}{\varepsilon}$$

where ε is a positive number supposed to be *small*, and functions $\alpha(\cdot)$, $\beta(\cdot)$ verify the same assumptions 5.1. Thus, the model [5.2] is written as:

$$\begin{cases} \dfrac{ds}{dt} &= D(S_{in} - s) - \mu_u(s)u - \mu_v(s)v \\[2mm] \dfrac{du}{dt} &= \mu_u(s)u - Du - \frac{1}{\epsilon}\left(\alpha(u,v)u - \beta(v)v\right) \\[2mm] \dfrac{dv}{dt} &= \mu_v(s)v - Dv + \frac{1}{\epsilon}\left(\alpha(u,v)u - \beta(v)v\right) \end{cases} \qquad [5.19]$$

It is convenient to write this dynamic by replacing the variables u and v by x and p, as we did in Chapter 3:

$$\begin{cases} \dfrac{ds}{dt} &= D(S_{in} - s) - \bar{\mu}(s,p)x \\[2mm] \dfrac{dx}{dt} &= \bar{\mu}(s,p)x - Dx \\[2mm] \dfrac{dp}{dt} &= \left(\mu_u(s) - \mu_v(s)\right)p(1-p) - \frac{1}{\epsilon}\left(\alpha(px, (1-p)x)p \\ & \quad - \beta((1-p)x)(1-p)\right) \end{cases} \qquad [5.20]$$

by defining:

$$\bar{\mu}(s,p) := p\,\mu_u(s) + (1-p)\,\mu_v(s)$$

Observe that this dynamic system is of the form:

$$\begin{cases} \dfrac{ds}{dt} &= f_s(s,x,p) \\[2mm] \dfrac{dx}{dt} &= f_x(s,x,p) \\[2mm] \dfrac{dp}{dt} &= \frac{1}{\epsilon}\left[\epsilon f_p(s,p) + g(x,p)\right] \end{cases}$$

where we posit:

$$g(x,p) := -\alpha(px, (1-p)x)p + \beta((1-p)x)(1-p)$$

When ϵ is small and the terms $f_s(s,x,p)$, $f_x(s,x,p)$ and $\epsilon f_p(s,p) + g(x,p)$ are of the same order of magnitude, the velocity $\frac{dp}{dt}$ is then very large compared to velocities $\frac{ds}{dt}$, $\frac{ds}{dt}$. Variables s and x can then be considered as almost constant and the approximation of the dynamics of variable p as "fast":

$$\frac{dp}{dt} = \frac{1}{\epsilon} g(x,p) \qquad\qquad [5.21]$$

where s is considered as a constant parameter (the term $\epsilon f_p(s,p)$ being negligible with regard to $g(x,p)$). If for any x, the differential equation [5.21] admits a unique globally asymptotically stable steady-state $\bar{p}(x)$, then this expression can be carried to the system [5.20] to obtain the "slow" approximation of the dynamics of the variables s and x:

$$\begin{cases} \dfrac{ds}{dt} &= D(S_{in} - s) - \mu(s,x)x \\[2mm] \dfrac{dx}{dt} &= \mu(s,x)x - Dx \end{cases} \qquad\qquad [5.22]$$

by defining:

$$\mu(s,x) = \bar{\mu}(s,\bar{p}(x))$$

This reduction technique (which consists of replacing ϵ by 0) is well known in physics under the name of *quasi-steady state approximation method*. At the mathematical level, the rigorous proof of the convergence of the solutions of the system [5.20] toward those of the reduced system [5.22] makes use of the theory of singular perturbations, which falls outside of the scope of this book.

PROPOSITION 5.3.– Under assumptions 5.1, there exists a unique function $\bar{p} : \mathbb{R}_+ \mapsto [0,1]$ C^1, strictly decreasing, such that $g(x,\bar{p}(x)) = 0$ for all $x > 0$. In addition, $\bar{p}(x)$ is the unique globally asymptotically stable steady-state of the scalar dynamic [5.21], for all $x > 0$.

Demonstration. For any $x > 0$, we have $g(x,0) = \beta(x) > 0$ and $F(x,1) = -\alpha(x,0) < 0$ (following assumptions 5.1). According to the intermediate value theorem, there therefore exists $\bar{p}(x) \in]0,1[$ such that $g(x,\bar{p}(x)) = 0$. Let us determine the partial derivatives of the function g:

$$\frac{\partial g}{\partial x} = -\left[(\alpha'_u(u)p + \alpha'_v(v)(1-p))p + \beta'(v)(1-p)^2\right]_{u=px,\,v=(1-p)x}$$

$$\frac{\partial g}{\partial p} = -\left[(\alpha'_u(u) - \alpha'_v(v))u + \alpha(u,v) + \frac{d}{dv}(\beta(v)v)\right]_{u=px,\,v=(1-p)x}$$

For $x > 0$, assumptions 5.1 guarantee $\frac{\partial g}{\partial x} < 0$ and $\frac{\partial g}{\partial p} < 0$. Thus, the function $p \mapsto g(x, p)$ is strictly decreasing, guaranteeing the uniqueness of the solution $\bar{p}(x)$ of $g(x, p) = 0$. According to the implicit function theorem, the function $\bar{p}$ is also differentiable for any $x > 0$ and its derivative is written as:

$$\bar{p}'(x) = -\frac{\frac{\partial g}{\partial x}(x, \bar{p}(x))}{\frac{\partial g}{\partial p}(x, \bar{p}(x))} < 0$$

The function $\bar{p}$ is thus C^1 on $\mathbb{R}_+ \setminus \{0\}$ and strictly decreasing.

Thereby, for all fixed $x > 0$, $\bar{p}(x)$ is the unique steady-state of the differential equation [5.21], and since $\frac{\partial g}{\partial p} < 0$ in all (x, p), it can be deduced that the steady-state $\bar{p}(x)$ is globally asymptotically stable for the scalar dynamics [5.21].

For example, for functions considered in [5.9], we obtain:

$$\bar{p}(x) = \frac{1}{1 + \frac{a}{b}x}$$

REMARK 5.1.– Following assumptions 5.1, we have:

$$\frac{\partial \mu}{\partial x}(s, x) = \frac{\partial \bar{\mu}}{\partial x}(s, x).\bar{p}'(x) = (\mu_u(s) - \mu_v(s)).\bar{p}'(x) < 0$$

and thus the model [5.22] for the total biomass x has a density-dependent growth, decreasing with respect to x, as it has been studied in Chapter 4 (for an isolated species).

EXERCISE 5.1.– It is considered that flocs contain only two individuals:

1) provide a simple expression for the function $\alpha(\cdot)$;

2) Then determine the corresponding function $\bar{p}$, for a constant function β.

EXERCISE 5.2.– Consider the case in which the terms D_u and D_v are distinct. We will denote λ_u, λ_v the break-even concentrations defined by $\mu_u(\lambda_u) = D_u$ and $\mu_u(\lambda_u) = D_u$;

1) write the slow approximation of the dynamics of variables s and x;

2) show that when $\lambda_u < \lambda_v$ there exists at most a positive steady-state, under the assumption that the function $x \mapsto x\bar{p}(x)$ is increasing;

3) show that in this case this steady-state is locally asymptotically stable;

4) show that when $\lambda_u > \lambda_v$, there is a possibility of having several positive steady-states;

5) when functions α and β verify $\alpha(0,0) = 0$ and $\beta(0) > 0$, show that washout steady-state is (locally) attractive as soon as $\lambda_u > S_{in}$ (independently of the value of λ_v).

5.2.6. *Consideration of several species*

When several species are in competition, we can similarly decompose the biomass of each species i into planktonic biomass u_i and attached biomass v_i (without differentiating the composition of flocs which can mix individuals from different species). The specific attachment functions α_i then depend (*a priori*) on all others quantities u_j, v_j since a free individual of species i can attach to free biomass or biomass with any species attached. Analogously, the specific detachment functions β_i depend *a priori* on all quantities v_j of biomass attached where an individual i could have attached. To simplify, it will be possible, for example, to assume that the α_i are functions of the total planktonic and attached biomass $u = \sum_j u_j$ and $v = \sum_v v_j$, and the β_i functions of v only, with the same assumptions [5.1]. The combinatorics of the possible specific cases makes the mathematical study much more complicated, but when the attachment and detachment velocities can be considered to be fast, the quasi-steady state approximation makes it possible to write a dynamic system for biomasses $x_i = u_i + v_i$ by expressing the terms u_i and v_i according to all the x_j on the "slow" manifold defined by the system of equations:

$$\alpha_i(u_1, \cdots, u_n, v_1, \cdots, v_n)u_i - \beta(v_1, \cdots, v_n)v_i = 0 \qquad i = 1 \cdots n$$

For example, by considering simple functions like we did in [5.9]:

$$\alpha_i(x_1, \cdots, x_n) = \sum_{j=1}^{n} a_{ij}x_j, \quad \beta_i = b_i$$

where parameters a_{ij} reflect how easily an individual of species i attaches to an individual of species j, the following expressions are obtained for the proportions $q_i = u_i/x_i$ on the slow manifold:

$$\bar{q}_i(x_1, \cdots, x_n) = \frac{1}{1 + \frac{1}{b_i}\sum_{j=1}^{n} a_{ij}x_j}$$

as in section 5.2.5 (under the assumption of fast attachments and detachments), and the reduced system is then written as:

$$\frac{dx_i}{dt} = \mu_i \left(S_{in} - \sum_{j=1}^{n} x_j, x_1, \cdots, x_n \right) x_i - Dx_i \qquad i = 1 \cdots n$$

by setting:

$$\mu_i(s, x) = \bar{q}_i(x_1, \cdots, x_n)\mu_{u_i}(s) + (1 - \bar{q}_i(x_1, \cdots, x_n))\mu_{v_i}(s)$$

as density-dependent growth functions, decreasing with respect to the x_i. This then corresponds exactly to the context of the study of Chapter 4, which shows that a coexistence between species is possible. It is thus concluded that a mechanism of (fast) attachment and detachment of biomass is a possible (theoretical) explanation for maintaining biodiversity in a chemostat.

5.3. The "predator-prey" relationship in the chemostat

5.3.1. *Introduction*

We are going to consider a situation in which several species of bacteria consume substrate continuously brought into a chemostat and where the bacteria of each species are the "prey" of a "predator" that is also itself a microorganism living in the same liquid medium as bacteria. A "predator" species may be common to several species of bacteria or not. A possible model of this situation is:

$$\begin{cases} \dfrac{ds}{dt} &= D(S_{in} - s) - \sum_i \mu_i(s, x_1, \cdots, x_n)x_i \\[2mm] \dfrac{dx_i}{dt} &= (\mu_i(s, x_1, \cdots, x_n) - D_{x_i})x_i - \sum_j \nu_{ij}(x_i)y_j \\[2mm] \dfrac{dy_j}{dt} &= \left(\sum_j \nu_{ij}(x_i) - D_{y_i}\right)y_j \end{cases} \qquad [5.23]$$

If in the model [5.23], we establish $y_i = 0$ (no predators of x_i), we recognize the density-dependent competition model (see section 2.3.3) where for x_i we have chosen units such that the yield term is equal to 1. In the terms:

$$\begin{cases} \dfrac{dx}{dt} &= \quad \cdots \quad -\sum_j \nu_{ij}(x_i)y_j \\[2mm] \dfrac{dy_j}{dt} &= \left(\sum_i \nu_{ij}(x_i) - D_{y_i}\right)y_j \end{cases}$$

we recognize in the first line the consumption of the "prey" (the resource x_i) by the predator and in the second the term of the growth associated with the consumption of all species followed by the term of removal. Here, we have not assumed that kinetics are density-dependent. A general theory of this type of model does not yet exist and in this book we will consider only very simple special cases. We will also assume that the μ_i are not density dependent and that all the D_{x_i} and D_{y_j} are equal to D.

5.3.2. *The substrate-bacteria-predator "chain"*

This is the case where there is a single species of bacteria consumed by a single predator. This is referred to as a "chain" because the pattern of interaction is a uni-dimensional "line" (Figure 5.4a). Hence:

$$\begin{cases} \dfrac{ds}{dt} &= D(S_{in} - s) - \mu(s)x \\[2mm] \dfrac{dx}{dt} &= (\mu(s) - D)x - \nu(x)y \\[2mm] \dfrac{dy}{dt} &= \big(\nu(x) - D\big)y \end{cases} \qquad [5.24]$$

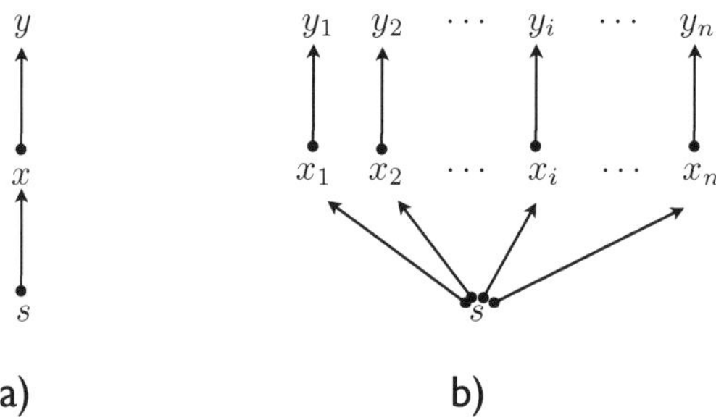

Figure 5.4. *Trophic "chain" a) and "network" b)*

We make the following assumptions:

HYPOTHESIS 5.2.–

1) the function $s \mapsto \mu(s)$ is continuous, differentiable, positive, zero in 0, strictly increasing and bounded:

$$0 \le \mu(s) < \mu_{max} \qquad \mu'(s) > 0$$

2) the function $x \mapsto \nu(x)$ is continuous, differentiable, positive, zero in 0, strictly increasing and bounded:

$$0 \le \nu(x) < \nu_{max} \qquad \nu'(s) > 0$$

From these assumptions, it follows that μ has an inverse function μ^{-1}, equal to zero at 0, strictly increasing and defined on $[0, \mu_{max}[$. Similarly ν has an inverse function ν^{-1}, equal to zero at 0, strictly increasing and defined on $[0, \nu_{max}[$. The sum $\mu^{-1} + \nu^{-1}$ is in turn a strictly increasing function that has an inverse function $\left(\mu^{-1} + \nu^{-1}\right)^{-1}$ defined on $[0, +\infty[$. Clearly, we have $\mu > \left(\mu^{-1} + \nu^{-1}\right)^{-1}$.

The change of variable:

$$\left(s, x, y\right) \mapsto \left(z = s + x + y,\ x, y\right)$$

gives:

$$\begin{cases} \dfrac{dz}{dt} &=& D(S_{in} - z) \\[2mm] \dfrac{dx}{dt} &=& \left(\mu(z - (x + y)) - D\right)x - \nu(x)y \\[2mm] \dfrac{dy}{dt} &=& \left(\nu(x) - D\right)y \end{cases} \qquad [5.25]$$

We are going to describe the operating diagram of this system. For this, we set S_{in} and make D decrease from $+\infty$ to 0.

5.3.2.1. *The case:* $\mu(S_{in}) < D$

The variable $z(t)$ tends to S_{in} thus $\mu(z - (x+y)) - D$ in the end becomes strictly negative and thereafter $\frac{dx}{dt}$. Therefore, $x(t)$ tends toward 0 which results in $\nu(x) - D$ being negative and thus $y(t)$ also tends to 0. In conclusion:

For $D > \mu(S_{in})$, the steady-state $\left(S_{in}, 0, 0\right)$ is globally stable: there is washout.

5.3.2.2. *The case:* $\left(\mu^{-1} + \nu^{-1}\right)^{-1}(S_{in}) < D < \mu(S_{in})$

If $D < \mu(S_{in})$, we get the steady-state:

$$E_1 = \left(S_{in}, (S_{in} - \mu^{-1}(D)), 0\right) = (z^*, x^*, 0)$$

In this steady-state the Jacobian matrix is:

$$J(E_1) = \left[\begin{array}{c|c|c} -D & 0 & 0 \\ \hline \mu'(z^* - x^*)x^* & -\mu'(z^* - x^*)x^* & -\mu'(z^* - x^*)x^* \\ \hline 0 & 0 & \nu(x^*) - D \end{array} \right] \qquad [5.26]$$

The steady-state is stable if and only if:

$$\nu(S_{in} - \mu^{-1}(D)) - D < 0$$

that is:

$$S_{in} < \mu^{-1}(D) + \nu^{-1}(D)$$

still:

$$(\mu^{-1} + \nu^{-1})^{-1}(S_{in}) < D$$

Now let us examine the possibility of a steady-state with $y^* > 0$. To cancel out the third equation we must have $x^* = \nu^{-1}(D)$. For the y^* determined by the second equation to be strictly positive it is necessary that:

$$\mu(S_{in} - \nu^{-1}(D)) - D > 0$$

that is:

$$S_{in} > \nu^{-1}(D) + \mu^{-1}(D)$$

still:

$$D < (\mu^{-1} + \nu^{-1})^{-1}(S_{in})$$

Therefore, when D "crosses" when decreasing the value $(\mu^{-1} + \nu^{-1})^{-1}(S_{in})$, the steady-state E_1 becomes unstable and a steady-state appears with biomass $y^* > 0$. What can be said of the stability of this steady-state?

The Jacobian matrix in $E_2 = (S_{in}, \nu^{-1}(D), y^*) = (z^*, x^*, y^*)$ is:

$$J(E_2) = \left[\begin{array}{c|c|c} -D & 0 & 0 \\ \hline \mu'x^* & -\mu'x^* + (\mu - D) - \nu'y^* & -\mu'x^* - \nu \\ \hline 0 & \nu'y^* & 0 \end{array}\right] \qquad [5.27]$$

where we have omitted to write the arguments of the functions μ, ν, μ', ν'. This matrix is stable if and only if:

$$-\mu'x^* + (\mu - D) - \nu'y^* < 0$$

For $D = (\mu^{-1} + \nu^{-1})^{-1}(S_{in})$, this quantity is strictly negative and remains so by continuity for smaller values close to $(\mu^{-1} + \nu^{-1})^{-1}(S_{in})$, it can, however, eventually become positive when D continues to decrease. When the steady-state becomes unstable periodic solutions do appear through Hopf bifurcation (see section 4.4.5 and [MAR 12]). Figure 5.5 shows the operating diagram of the model:

$$\begin{cases} \dfrac{ds}{dt} &= D(S_{in} - s) - \dfrac{s}{0.2 + s}x \\ \dfrac{dx}{dt} &= (\dfrac{s}{0.2 + s} - D)x - \dfrac{x}{1 + x}y \\ \dfrac{dy}{dt} &= (\dfrac{x}{1 + x} - D)y \end{cases} \qquad [5.28]$$

The graphs of μ and $(\mu^{-1} + \nu^{-1})^{-1}$ have been drawn based on their analytical expression. The red curve separating the "Steady-state with predator" region from the "periodic solutions" region is the curve:

$$\mu'(s^*)x^* + (\mu(s^*) - D) - \nu'(x^*)y^*$$

where s^*, x^*, y^* are determined according to S_{in} and D.

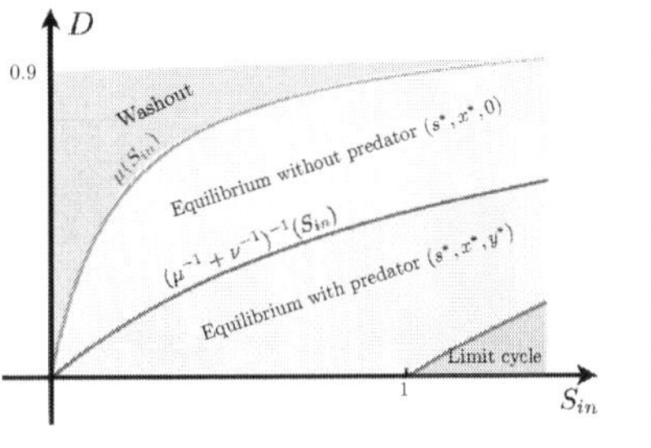

Figure 5.5. *The operating diagram of [5.28]. For a color version of this figure, see www.iste.co.uk/harmand/chemostat.zip*

5.3.3. *The substrate-bacteria-predators trophic network*

The expression "trophic network", as opposed to "trophic chain", is employed when the interactions graph is no longer linear: for example when several different species of bacteria in competition are, in turn, subject to predations (see Figure 5.4(b)). The interesting point in the presence of predators in competition models is that it can allow for the coexistence of several species whereas in their absence there is no coexistence. In effect, let such a model be:

$$\begin{cases} \dfrac{ds}{dt} &= D(S_{in} - s) - \sum_{i=1}^{n} \mu_i(s)x_i \\ \dfrac{dx_i}{dt} &= (\mu_i(s) - D)x_i - \nu_i(x_i)y_i \quad i = 1, \cdots, n \\ \dfrac{dy_i}{dt} &= (\nu_i(x_i) - D)y_i \quad i = 1, \cdots, n \end{cases} \qquad [5.29]$$

In this model, a predator is associated (concentration y_i) with each species of bacteria (concentration x_i). For every μ_i and ν_i, we make the assumptions 5.2. The steady-states are given by the solutions of:

$$\begin{cases} 0 &= D(S_{in} - s) - \sum_{i=1}^{n} \mu_i(s)x_i \\ 0 &= (\mu_i(s) - D)x_i - \nu_i(x_i)y_i \quad i = 1, \cdots, n \\ 0 &= \big(K_i \nu_i(x_i) - D\big)y_i \quad i = 1, \cdots, n \end{cases} \qquad [5.30]$$

From the third series of equations:

$$(\nu_i(x_i) - D)x_i \quad i = 1, \cdots, n \qquad [5.31]$$

We determine all non-zero x_i^* that carried to the first equation give us:

$$D(S_{in} - s) = \sum_{i=1}^{n} \mu_i(s)x_i^* \qquad [5.32]$$

which always has a unique solution s^* since the left member is a decreasing function of DS_{in} at 0, whereas the right one is increasing from 0. The second series of equations gives us the y_i^* when solving:

$$(\mu_i(s^*) - D)x_i^* - \nu_i(x_i^*)y_i^* = 0 \qquad [5.33]$$

However, these last equations have positive solutions only if $\mu_i(s^*) - D > 0$. A first condition is thus that $D < \mu_i(S_{in})$. Then, all depends of the value of s^*. As with the example of the trophic chain, it can be shown that for D being small enough, all $\mu_i(s^*)$ are larger than D. In effect, we have $x_i^* = \nu_i^{-1}(D)$ and thus $s^* = S_{in} - \sum_i \nu_I^{-1}(D) - \sum_i y_i^*$. It is thus necessary that:

$$D < \mu_i\big(S_{in} - \sum_i \nu_i^{-1}(D)\big)$$

that is, for all i:

$$\big(\mu_i^{-1} + \sum_i \nu_i^{-1}\big)(D) < S_{in}$$

and thus:

$$D < \min_i \left(\mu_i^{-1} + \sum_i \nu_i^{-1} \right)^{-1} (S_{in})$$

What can be said of the stability of this equilibrium?

Note that the system [5.29] can be rewritten as:

$$\begin{cases} \dfrac{ds}{dt} = D(S_{in} - s) - \displaystyle\sum_{i=1}^{n} \mu_i(s)x_i \\[2mm] \begin{cases} \dfrac{dx_i}{dt} = (\mu_i(s) - D)x_i - \nu_i(x_i)y_i \\[2mm] \dfrac{dy_i}{dt} = (\nu_i(x_i) - D)y_i \end{cases} \quad i = 1, \cdots, n \end{cases} \tag{5.34}$$

Whose Jacobian matrix is:

$$\begin{pmatrix} -A & -\mu_1 & 0 & -\mu_i & 0 & -\mu_n & 0 \\ \cdots & \cdots & \cdots & \cdots & \cdots & \cdots & \cdots \\ \cdots & \cdots & \cdots & \cdots & \cdots & \cdots & \cdots \\ \mu_i'x_i^* & 0 & 0 & (\mu_i - D) - \nu_i'y_i^* & -\nu_i & 0 & 0 \\ 0 & 0 & 0 & \nu_i'y_i^* & 0 & 0 & 0 \\ \cdots & \cdots & \cdots & \cdots & \cdots & \cdots & \cdots \\ \cdots & \cdots & \cdots & \cdots & \cdots & \cdots & \cdots \\ \mu_n'x_n^* & 0 & 0 & 0 & 0 & (\mu_n - D) - \nu_n'y_n^* & -\nu_n \\ 0 & 0 & 0 & 0 & 0 & \nu_n'y_n^* & 0 \end{pmatrix}$$

with $A = \left(D + \sum_{i=1}^{n} \mu_i'x_{ie} \right)$ and where the function arguments are omitted.

This matrix is not simple to study unless we make the assumption that all the $\mu_i'(s^*)$ are equal to zero (or at least negligible compared to the other terms). Such an assumption is not totally unrealistic if we have a large concentration at steady-state compared with the saturation concentration of μ_i. Then the matrix is diagonal by blocks, the eigenvalues are $\lambda_0 = -(D + \sum_{i=1}^{n} \mu_i'x_i^*)$ and the two roots of the equations:

$$\lambda^2 - (\mu_i - D) - \nu_i'y_i^*)\lambda + \nu_i'\nu_i y_i^* = 0 \tag{5.35}$$

It all therefore depends on the sign of:

$$Tr_i = (\mu_i - D) - \nu_i'y_i^*$$

which can be positive or negative depending on the assumptions made.

We will not continue this discussion. We will retain only that it is possible that the presence of predators of each species of bacteria, for example, of specific viruses, allows coexistence.

5.3.4. *Comparison to experimental data*

In the article [TSU 72] *Predator-Prey Interactions of* Dictyostelium discoideum *and* Escherichia coli *in Continuous Culture*, the authors consider a culture of *Escherichia coli* bacteria in a chemostat where the substrate is glucose. In the same chemostat, they add a *Dictyostelium discoideum* amoeba that feeds on *Escherichia coli* but that is known to not absorb glucose.

They effectively observe, by varying the inflow in the chemostat, the possibility of steady-state without predator, steady-state with predator and oscillations finally. Figure 5.6 shows the evolution over time of glucose (on top), of *Escherichia coli* (in the middle) and of *Dictyostelium discoideum* concentrations (at the bottom) based on data from [TSU 72] in the case where oscillations are observed.

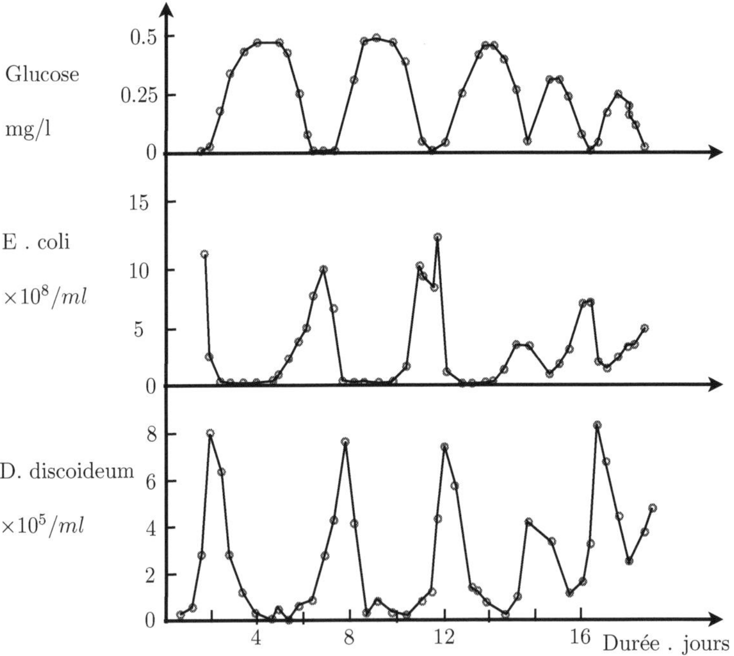

Figure 5.6. *According to [TSU 72]. For a color version of this figure, see www.iste.co.uk/harmand/chemostat.zip*

The authors of these experiences propose the following model:

$$\begin{cases} \dfrac{ds}{dt} &= D(S_{in} - s) - 0.033 \,\dfrac{0.25\,s}{0.0005 + s}x \\[2mm] \dfrac{dx}{dt} &= \left(\dfrac{0.25\,s}{0.0005 + x} - D\right)x - 1.4\dfrac{0.24\,x}{4 + x}y \\[2mm] \dfrac{dy}{dt} &= \left(\dfrac{0.24\,x}{4 + x} - D\right)y \end{cases} \qquad [5.36]$$

The units are:

– time: hour;

– glucose (substrate) s: mg per ml;

– *escherichia coli* x : 10^9 cells per ml;

– *dictyostelium discoideum* y : 10^5 cells per ml.

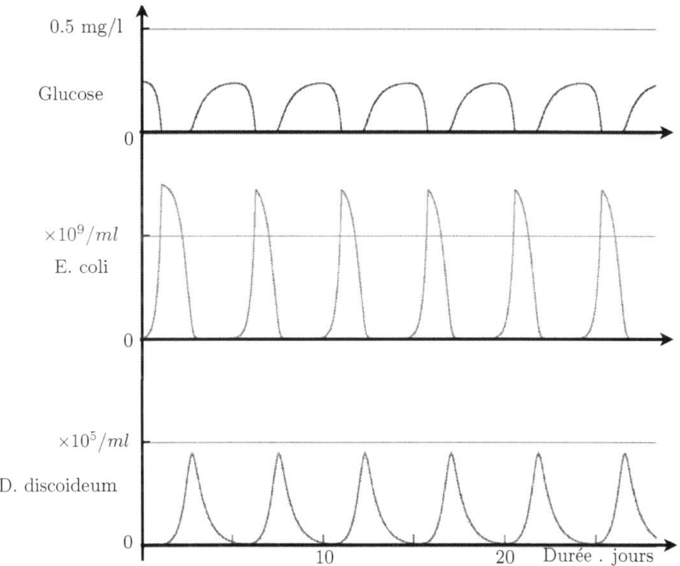

Figure 5.7. *A simulation of [5.36]:* $S_{in} = 0.5\ D = 0.0625$. *For a color version of this figure, see www.iste.co.uk/harmand/chemostat.zip*

The values of the constants of the two Monod functions are experimentally determined with separate cultures, and then the entire model is simulated, which yields Figure 5.7 in which we can correctly observe oscillations comparable to those empirically observed. We recommend reading [TSU 72] (which is available online for free) where the model can be seen fitting to the data and in which more information can be found about the details of the experiments.

5.4. Bibliographic notes

Attachment and detachment phenomena of bacteria, whether in biofilms on a support [COS 95, IWA 06] or in the form of aggregates or flocs [THO 99] are well known and frequently observed in bacterial growth. Nevertheless, it is only relatively recently that they have been explicitly taken into account in chemostat-based mathematical models. The Freter model [FRE 83, JON 03], proposed in the 1980s as a functional model of the intestine bacterial ecosystem, is one of the very first to explicitly distinguish *planktonic* biomass from attached biomass. This model considers specific attachment and detachment terms and has been mathematically studied in a spatialized form by introducing advection and diffusion terms [BAL 08].

Several works in the biomathematical literature consider extensions to the chemostat model spatialized with (fixed) attachment on a wall by [BAL 99, JON 03, STE 00]. In general, flocculation models describe the dynamics of the distribution of floc sizes [THO 99] and their influence on growth dynamics [HAE 07], but comparatively there are relatively few studies of simplified models that only distinguish two biomass compartments: planktonic and attached. In [HAE 08], it is shown for such models that total biomass growth follows a density-dependent distribution, under the assumption that attachment and detachment velocities are large compared to biological terms. This is in accordance with experimental observations that have showed that the kinetics of processes with attached biomass are better represented by ratio-dependent [HAR 07] expressions.

The majority of the models in the literature consider explicit attachment and detachment term expressions. The presentation adopted in this chapter, which does not particularize the terms $\alpha(\cdot)$ and $\beta(\cdot)$, thus proposes a more general framework, which namely includes the models [JON 03, PIL 99, TAN 97]. In every case, the assumptions concerning faster growth and higher planktonic bacteria removal rate are justified by experimental observations [HEF 09]. It should be observed that attachment and detachment velocities can be of a very variable order of magnitude, according to procedures and operating conditions [BER 95], justifying the fact of either considering the global model [5.2], or the reduced model as we did in section 5.2.5.

Nevertheless, due to concerns of exposition and simplification, we have considered a number of restrictions (in particular: the expressions α and β do not depend on the resource s, the removal terms D_u, D_v are identical). The interested reader is encouraged to consult more technical expansions [FEK 13b, FEK 16] (in particular, considering the terms D_u and D_v as being distinct brings multi-stability and an enrichment of non-intuitive solution behaviors).

In 1972, [TSU 72] experimentally demonstrated the presence of oscillations in a resource-consumer-predator relation in a chemostat. They proposed a model of the type discussed in section 5.3.2 which properly adjusted to the observed data. In 1989, [KUA 89] mathematically demonstrated that this type of model had a limit cycle. At the same date, [WOL 89] proposed a comprehensive analysis of the type of networks referred to in section 5.3.3 under the particular assumption that functions ν_i are linear.

Appendices

Appendix 1

Differential Equations

The purpose of this appendix is to present the results on differential equations that have been used throughout this book. In order to understand this appendix, the reader should be familiar with concepts of linear algebra (vectors, matrices, eigenvalues and eigenvectors) as well as with the basic concepts of analysis (continuity and partial derivative) and topology (open sets, neighborhoods, etc.). There are countless books on the theory of differential equations. Among them, we can cite: [ANO 88, ARN 74, DEM 06, GUC 13, HAH 67, HAL 69, HIR 74, LEF 63, NEM 60, PER 13]. We invite the reader who wishes to learn more about differential equations to consult any of them.

A1.1. Definitions, notations and fundamental theorems

A1.1.1. *Systems of differential equations*

It is assumed that the state of a system, at a given time t, can be represented by a vector $x(t) = (x_1(t), \ldots, x_n(t))$ of a state space $\mathcal{D}$ which is a subset of $\mathbb{R}^n$ and that the function $t \mapsto x(t)$ is the solution of a differential system:

$$\frac{dx}{dt} = f(x) \tag{A1.1}$$

where $f : x \in \mathcal{D} \to f(x) \in \mathbb{R}^n$ is a continuous function. The subset $\mathcal{D}$ is the domain of definition of the differential system [A1.1]. A function $t \mapsto x(t)$ is said to be a solution of [A1.1], if this function is defined on an interval $I \subseteq \mathbb{R}$, it takes its values in $\mathcal{D}$, it is continuously differentiable on I, and for any $t \in I$, we have $\frac{dx}{dt}(t) = f(x(t))$.

Let us clarify the notations: $t \mapsto \frac{dx}{dt}(t)$ is the tangent vector of $t \mapsto x(t)$. For all t, this vector is equal to $f(x(t))$ which is itself a vector of $\mathbb{R}^n$:

$$\frac{dx}{dt} = \begin{bmatrix} \frac{dx_1}{dt} \\ \vdots \\ \frac{dx_n}{dt} \end{bmatrix}, \qquad x = (x_1, \ldots, x_n) \in \mathcal{D} \mapsto f(x) = \begin{bmatrix} f_1(x_1, \cdots, x_n) \\ \vdots \\ f_n(x_1, \cdots, x_n) \end{bmatrix}$$

The mapping $x \in \mathcal{D} \mapsto f(x) \in \mathbb{R}^n$ is called a vector field on $\mathcal{D}$. The evolution of the system is described by [A1.1] which is a system of n differential equations:

$$\frac{dx_i}{dt} = f_i(x_1, \cdots, x_n), \qquad i = 1 \cdots n$$

As a result, a solution of [A1.1] is a continuously differentiable function:

$$t \in I \mapsto x(t) = (x_1(t), \ldots, x_n(t)) \in \mathcal{D}$$

such that for any $t \in I$, we have:

$$\frac{dx_i}{dt}(t) = f_i(x_1(t), \cdots, x_n(t)), \qquad i = 1 \cdots n$$

To solve or *to integrate* a differential system means finding all of the solutions thereof. It will be seen that a differential system has an infinite number of solutions and that if an (initial) condition $x(0) = x_0 \in D$ is fixed then the solution is unique.

DEFINITION A1.1.– *Let* $t \in I \mapsto x(t) \in \mathcal{D}$ *be a solution of* [A1.1]. *The subset* $\{x(t) : t \in I\}$ *of* $\mathcal{D}$ *is called the* orbit *(or the trajectory) of the solution $x(t)$.*

Among the solutions of a differential system can be distinguished the *equilibrium points* (or *steady states*, or *fixed points*, or *singular points*) that play an important role in the description of the solutions of the system.

DEFINITION A1.2.– *A point* $x^* \in D$ *is said to be an equilibrium point of* [A1.1], *if* $f(x^*) = 0$, *or still, equivalently, if the function $x(t) = x^*$ is a solution of* [A1.1].

If x^* is a equilibrium point, then the orbit associated with the solution $x(t) = x^*$ is equal to $\{x^*\}$. If x^* is not an equilibrium point (namely that $f(x^*) \neq 0$, it is then said that x^* is a *regular point*, as opposed to *singular point*), and if $t \mapsto x(t)$ is a solution of [A1.1] that passes through this point, in other words for which there is a time t^* such that $x^* = x(t^*)$ then the orbit corresponding to the solution is a curve parameterized by t which is tangent to the vector $f(x^*)$ at point x^*. From a geometrical perspective, an orbit is thus tangent at each of its points to the vector field $f : \mathcal{D} \to \mathbb{R}^n$.

In the differential systems encountered in population dynamics, the state variables $x_i(t)$ have a physical meaning (for example they are concentrations or densities). As a result, the state space (or phase space) is the positive orthant defined by:

$$\mathbb{R}^n_+ = \{x \in \mathbb{R}^n : x_i \geq 0, i = 1\ldots n\}$$

or still a subset of this set if, for example, some state variables cannot take all positive values.

EXAMPLE A1.1.– The competition model [3.1] of Chapter 3 of several species in the chemostat is a differential system defined in the orthant $\mathcal{D} = \mathbb{R}^{n+1}_+$. The state of the system at time t is represented by the vector $(s(t), x_1(t), \ldots, x_n(t))$ where $s(t)$ is the substrate concentration and $x_i(t)$, $i = 1 \cdots n$, that of species. The vector field associated with this system is given by:

$$(s, x_1, \ldots, x_n) \in \mathcal{D} \mapsto f(s, x_1, \ldots, x_n) = \begin{bmatrix} D(S_{in} - s) - \sum_{i=1}^{n} \mu_i(s)x_i \\ (\mu_1(s) - D)x_1 \\ \vdots \\ (\mu_n(s) - D)x_n \end{bmatrix}$$

EXERCISE A1.1.– Consider the differential system:

$$\begin{aligned} \frac{ds}{dt} &= D(S_{in} - s) - p(s)x \\ \frac{dx}{dt} &= (q(s) - D_1)x \end{aligned} \tag{A1.2}$$

defined on the domain $\mathcal{D} = \{(s, x) \in \mathbb{R}^2 : s \geq 0, x \geq 0\}$, where $s \mapsto p(s)$ and $s \mapsto q(s)$ are positive functions for $s > 0$ and equal to zero at 0. When $p(s) = q(s)$, this is the *minimal model* of the chemostat considered in Chapter 2. Show that the system [A1.2] admits as equilibrium points $E_0 = (S_{in}, 0)$ or $E_1 = (s^*, x^*)$ where s^* is defined by $q(s^*) = D_1$ and $x^* = \pi(s^*)$ with:

$$\pi(s) = \frac{D(S_{in} - s)}{p(s)}$$

The washout steady-state E_0 exists. Show that the (positive) steady-state E_1 exists if and only if $S_{in} > s^*$. If the equation $q(s) = D_1$ admits several solutions, there may be several positive steady-states of the E_1 type.

The simplest example of a differential system is that of the scalar linear differential equation:

$$\frac{dx}{dt} = ax \tag{A1.3}$$

defined on $\mathcal{D} = \mathbb{R}$, in which a is a real constant. This differential equation can be integrated, as shown in the following proposition.

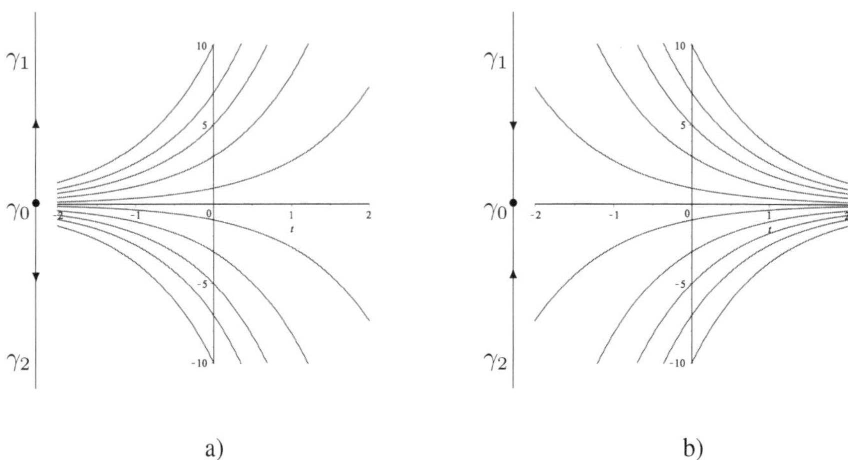

a)   b)

Figure A1.1. *Solutions of [A1.3] with initial conditions* $x_0 = 0, \pm 1, \pm 3, \pm 5, \pm 7, \pm 10,$ *with: a)* $a = 1$ *(exponential growth); b)* $a = -1$ *(exponential decay). For a color version of this figure, see www.iste.co.uk/harmand/chemostat.zip*

PROPOSITION A1.1.– The function $t \mapsto x_0 e^{at}$ is a solution of [A1.3]. It is the only solution that is equal to x_0 for $t = 0$.

Demonstration. Let us denote $x(t) = x_0 e^{at}$. It follows that: $\frac{dx}{dt}(t) = x_0 a e^{at} = ax(t)$. Consequently, $x(t) = x_0 e^{at}$ is a solution of [A1.3] such that $x(0) = x_0$. Let $y(t)$ be another solution of [A1.3] verifying $y(0) = x_0$. It follows that:

$$\frac{d}{dt}\left(y(t)e^{-at}\right) = \frac{d}{dt}(y(t))e^{-at} + y(t)\frac{d}{dt}\left(e^{-at}\right) = ay(t)e^{-at} - ay(t)e^{-at} = 0$$

As a result, $y(t)e^{-at}$ is a constant K and therefore $y(t) = Ke^{at}$. From the condition $y(0) = x_0$, it can be deduced that $K = x_0$.

The solutions of this differential equation are represented in Figure A1.1. In this figure, the orbits are represented on the vertical axes. Note that a same orbit is associated with an infinite number of solutions. If $a \neq 0$, there are three distinct orbits, $\gamma_0 = \{0\}$, which is a equilibrium point, $\gamma_1 = \{x \in \mathbb{R} : x > 0\}$ and $\gamma_2 = \{x \in \mathbb{R} : x < 0\}$. The arrows on the orbits indicate their travel direction when time t increases. The sign of a gives the asymptotic behavior (that is the behavior when t tends to infinity) of the solution $x(t) = x_0 e^{at}$.

PROPOSITION A1.2.– If $a > 0$, then $\lim\limits_{t \to +\infty} x_0 e^{at} = +\infty$ (if $x_0 > 0$).

If $a < 0$, then $\lim\limits_{t \to +\infty} x_0 e^{at} = 0$.

A1.1.2. *Existence and uniqueness of solutions*

What has been demonstrated for the scalar differential equation [A1.3], namely, the existence and uniqueness of the solution of fixed initial condition (see proposition A1.1), remains true for the differential system [A1.1], with some subtleties, such as the fact that the solutions are not necessarily defined on all $\mathbb{R}$, as in the case of [A1.3]. It is shown that if the function f is fairly regular (for example if it admits continuous partial derivatives), then the sequence $(x_k)_{k \geq 0}$ defined from an initial condition $x_0 \in \mathcal{D}$ by the recursion formula:

$$x_{k+1} = x_k + h f(x_k) \tag{A1.4}$$

where h is a very small time step, has a limit (in a sense that will be specified) when h tends to 0. This limit is the unique solution of [A1.1] which is equal to x_0 for $t = 0$.

The formula [A1.4] is called the Euler method (or the Euler algorithm) for the computation of the solutions of the system differential [A1.1]. Note that the sequence $(x_k)_{k \geq 0}$ is defined as long as x_k remains in the domain $\mathcal{D}$. More specifically, for $t > 0$, the piecewise constant function $t \mapsto x(t, h)$ is considered, which is defined by:

$$x(t, h) = x_k \quad \text{for } kh \leq t < (k+1)h \tag{A1.5}$$

If t is restricted to an interval $[0, \delta)$ with δ small enough, there is certainty that all the x_k are well defined and remain in $\mathcal{D}$ so that this function is well defined and remains in $\mathcal{D}$. Note that this function depends on the step h. When h tends to 0, it is shown (for details refer to [DEM 06], section V.2, pp. 131–141) that the function defined by [A1.5] has a limit:

$$x(t) = \lim\limits_{h \to 0} x(t, h)$$

and that this limit is the solution of [A1.1] such that $x(0) = x_0$. To demonstrate that the solution is also defined for $t < 0$, it suffices to take $h < 0$ and to perform the same construction. We thus have the following result.

THEOREM A1.1.– If f admits continuous partial derivatives in $\mathcal{D}$, then for all $x_0 \in \mathcal{D}$, there exists $\delta > 0$ and a unique solution of the differential system [A1.1] of initial condition $x(0) = x_0$, defined on $(-\delta, \delta)$.

EXERCISE A1.2.– The purpose of this exercise is to demonstrate that when $f(x) = ax$, the Euler method [A1.4] properly converges to the solution $x(t) = x_0 e^{at}$ of the scalar equation [A1.3] that we have obtained in proposition A1.1. Show that the function $x(t, h)$ defined by [A1.5] verifies:

$$(1 + ah)^{t/h-1} \, x_0 < x(t, h) \leq (1 + ah)^{t/h} \, x_0 \qquad [A1.6]$$

Show that:

$$e^{at} = \lim_{h \to 0} (1 + ah)^{t/h} = \lim_{h \to 0} (1 + ah)^{t/h-1} \qquad [A1.7]$$

As a result, based on [A1.6] and [A1.7], it can be deduced that $\lim_{h \to 0} x(t, h) = x_0 e^{at}$.

It is shown that the error occurring when approximating the solution with the Euler method is of the order of h. It should be noted that mathematicians and computer scientists have invented methods of greater orders than h, therefore more effective, in the sense that they use steps h not too small. As a result, they make it possible to obtain the solution, with the same accuracy, and with fewer iterations than Euler's method [DEM 06].

EXAMPLE A1.2.– The uniqueness of the solution with given initial condition is not guaranteed if the function f is continuous only, as shown in the example of the vector field $x \mapsto 3x^{2/3}$, continuous on $\mathbb{R}$ (but not differentiable) at 0. The differential equation:

$$\frac{dx}{dt} = 3x^{2/3} \qquad [A1.8]$$

associated with this vector field does not have the uniqueness of solutions, verifying $x(0) = 0$. For example, the functions $x_1(t) = 0$ and $x_2(t) = t^3$ are solutions of [A1.8], such that $x_1(0) = x_2(0) = 0$. There is an infinite number of solutions passing through 0 at time 0 (see exercise A1.3). Note that in this example, a solution can reach a point of equilibrium in finite time. For example, the function $x(t)$ defined by:

$$(t) = \begin{cases} (t - 1)^3 & \text{if } t < 1 \\ 0 & \text{if } t \geq 1 \end{cases}$$

is a solution of [A1.8] such that $x(0) = -1$ and that reaches equilibrium 0 at time 1. This property may not occur for a differential system with uniqueness of the solution going through a point.

EXERCISE A1.3.– Show that for all a and b such that $-\infty \leq a \leq 0 \leq b \leq +\infty$ the function $x(t)$ defined by:

$$x(t) = \begin{cases} (t-a)^3 & \text{if } t < a \\ 0 & \text{if } a \leq t \leq b \\ (t-b)^3 & \text{if } t > b \end{cases}$$

is a solution of [A1.8] such $x(0) = 0$.

The local solution whose existence is guaranteed by theorem A1.1 is defined on a maximal interval $I(x_0)$ of ends $t_{min}(x_0)$ and $t_{max}(x_0)$, depending on the initial point $x_0 \in D$. The maximum solution is denoted by $x(t, x_0)$ to remind its dependence according to x_0. We have the following result.

PROPOSITION A1.3.– The function $x(t, x_0)$ is continuous with respect to x_0. For all $s \in I(x_0)$ and all $t \in I(x(s, x_0))$, we have $t + s \in I(x_0)$ and $x(t, x(s, x_0)) = x(t + s, x_0)$.

Demonstration. Concerning continuity, refer to [HIR 74]. To demonstrate the formula $x(t, x(s, x_0)) = x(t+s, x_0)$, we denote $y(t) = x(t, x(s, x_0))$ and $z(t) = x(t+s, x_0)$. These two functions are solutions of [A1.1] such that:

$$y(0) = x(0, x(s, x_0)) = x(s, x_0), \text{ and } z(0) = x(s, x_0)$$

Since they are equal for $t = 0$, they are for every t.

A1.1.3. *Invariant sets*

Let us recall that the set:

$$\gamma(x_0) = \{x(t, x_0) : t \in I(x_0)\} \subset D$$

is the orbit (or the trajectory) passing by x_0.

DEFINITION A1.3.– *A set $A \subset D$ is said to be invariant by the system [A1.1] if for all $x_0 \in A$, we have $\gamma(x_0) \subset A$, which is equivalent to saying that for all $t \in I(x_0)$, we have $x(t, x_0) \in A$. If A verifies the property $x(t, x_0) \in A$ for any $x_0 \in A$ and any $t > 0$, it is then said that A is positively invariant.*

When t tends toward the end $t_{max}(x_0)$ of the maximal definition interval, either $x(t, x_0)$ tends to a boundary point of D, or tends to infinity, or still the solution always remains in D and in this case, the solution is defined for all $t \geq 0$, in other words the end $t_{max}(x_0)$ of the maximal definition interval is infinite. The same property is held for the end $t_{min}(x_0)$. Consequently, the trajectory $(t, x(t, x_0))$ leaves every compact set included in $\mathbb{R} \times D$. For more details, refer to [ARN 74]. This result is used in practice in the following way.

THEOREM A1.2.– If a solution $x(t, x_0)$ is positively bounded, that is if it enters a bounded and closed region of the domain D and remains therein for any $t \geq t_0$, then $t_{max}(x_0) = +\infty$. That is to say, $x(t, x_0)$ is defined for every $t \in [0, +\infty)$.

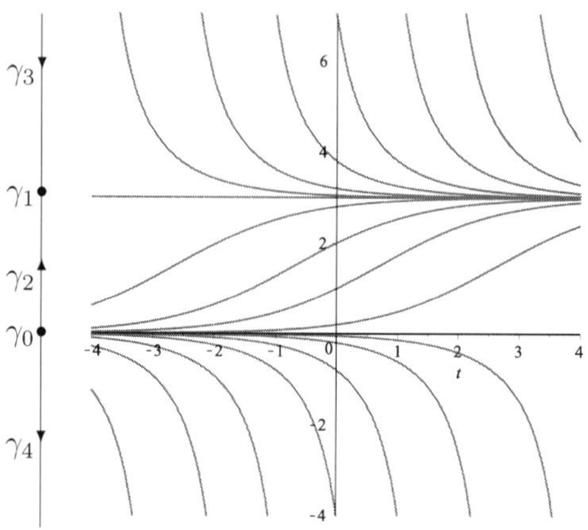

Figure A1.2. *Solutions of [A1.9], with cases $r = 1$ and $K = 3$. For a color version of this figure, see www.iste.co.uk/harmand/chemostat.zip*

EXAMPLE A1.3.– Consider the differential equation known as the logistic equation:

$$\frac{dx}{dt} = rx(1 - x/K) \tag{A1.9}$$

where r and K are positive parameters. Note that $x = 0$ and $x = K$ are the only equilibrium points. To solve this equation, we define $y = 1/x$. It follows that:

$$\frac{dy}{dt} = -\frac{1}{x^2}\frac{dx}{dt} = \frac{r}{K} - \frac{r}{x} = \frac{r}{K} - ry$$

Therefore, $y(t) = ce^{-rt} + \frac{r}{K}$ where c is a constant which is determined with the condition $y(0) = y_0$. We thus have: $c = y_0 - \frac{r}{K}$, from which it can be deduced that:

$$y(t) = y_0 e^{-rt} + \frac{1}{K}\left(1 - e^{-rt}\right)$$

Since $y_0 = 1/x_0$, with $x(0) = x_0$, we thus have:

$$x(t, x_0) = \frac{x_0 e^{rt}}{1 + \frac{x_0}{K}\left(e^{rt} - 1\right)}$$

If $x_0 \in [0K]$, this solution is defined for every $t \in \mathbb{R}$. Note that $x(t, 0) = 0$ and $x(t, K) = K$ for any t which was known already because $x_0 = 0$ and $x_0 = K$ are equilibrium points. If $x_0 > K$, the solution is defined on $(t_0, +\infty)$ and it tends to $+\infty$ when t tends to t_0, with:

$$t_0 = \ln \left(\frac{x_0 - K}{x_0} \right)$$

From a mathematical perspective, it can obviously be considered that the logistic equation is defined on the whole set $\mathbb{R}$. If $x_0 < 0$, the solution is defined on $(-\infty, t_0)$ and it tends to $-\infty$ when t tends to t_0, with t_0 previously defined. The solutions of this differential equation are represented in Figure A1.2. In this figure, the orbits are represented on the vertical axis. A same orbit is associated with an infinite number of solutions. There are five distinct orbits, $\gamma_0 = \{0\}$ and $\gamma_1 = \{K\}$ are equilibrium points, as well as $\gamma_2 = \{x \in \mathbb{R} : 0 < x < K\}$, $\gamma_3 = \{x \in \mathbb{R} : x > K\}$ and $\gamma_4 = \{x \in \mathbb{R} : x < 0\}$. The arrows on the orbits indicate their travel direction when time t increases.

EXERCISE A1.4.– Consider the differential system [A1.2]. Show that solutions are positively bounded and thus defined for all $t \geq 0$.

EXERCISE A1.5.– A non-singular point x_0 is said to be *periodic*, if there exists $T > 0$ such that $x(T, x_0) = x_0$ and $x(t, x_0) \neq x_0$ for all $t \in]0, T[$. Show the corresponding solution $t \mapsto x(t, x_0)$ is T-periodical.

PROPOSITION A1.4.– If $x(t)$ is a solution of [A1.1] such that $\lim_{t \to +\infty} x(t) = x^*$, then x^* is an equilibrium point. The same property holds if $\lim_{t \to -\infty} x(t)$ exists.

Demonstration. Denote by $y(t, y_0)$ the solution of the equation such that $y(0) = y_0$. We have (thus is continuity with respect to the initial conditions):

$$\lim_{s \to +\infty} y(t, x(s)) = y(t, x^*)$$

Moreover, since $y(t, x(s)) = x(t + s)$ (see proposition A1.3), we have:

$$\lim_{s \to +\infty} y(t, x(s)) = \lim_{s \to +\infty} x(t + s) = x^*$$

Therefore, we have $y(t, x^*) = x^*$ for every t and therefore x^* is an equilibrium point.

We have the following result, which shows that a solution of a differential system verifies an integral equation.

PROPOSITION A1.5.– If $t \in I \mapsto x(t)$ is a solution of [A1.1], then for all t_0 and t_1 in I, it follows that:

$$x(t_1) = x(t_0) + \int_{t_0}^{t_1} f(x(t))dt \qquad [\text{A1.10}]$$

Demonstration. For all t between t_0 and t_1, we have $\frac{dx}{dt}(t) = f(x(t))$. Integrating this expression between t_0 and t_1 yields that:

$$x(t_1) - x(t_0) = \int_{t_0}^{t_1} \frac{dx}{dt}(t)dt = \int_{t_0}^{t_1} f(x(t))dt$$

Consequently, $x(t_1) = x(t_0) + \int_{t_0}^{t_1} f(x(t))dt$, which demonstrates [A1.10].

The next exercise proposes another proof of proposition A1.4 based on formula [A1.10].

EXERCISE A1.6.– Let $x(t)$ be a solution of [A1.1] such that $\lim_{t \to +\infty} x(t) = x^*$. Show that if there exists i such that $f_i(x^*) \neq 0$, then $x_i(t)$ tends to infinity, which contradicts the existence of the limit x^*. As a result, $f(x^*) = 0$, which means that x^* is an equilibrium point.

A1.2. Theory of stability

A1.2.1. *Equilibria stability*

We consider the system [A1.1] that is again written here as:

$$\frac{dx}{dt} = f(x) \qquad [\text{A1.11}]$$

defined on a domain D of $\mathbb{R}^n$. It is assumed that f admits continuous partial derivatives.

DEFINITION A1.4.– *An equilibrium point x^* of [A1.11] is said to be*:

– *stable*, if for $\varepsilon > 0$, there exists $\delta > 0$ such that for any initial condition x_0, it yields that:

$$\|x_0 - x^*\| < \delta \implies \forall t > 0 \ \|x(t, x_0) - x^*\| < \varepsilon$$

which means that solutions remain as close as desired to x^* provided that they are close enough at time 0;

– *unstable*, if it is not stable;

– *attractive*, if there exists $\eta > 0$ such that for any initial condition x_0, we have:

$$\|x_0 - x^*\| < \eta \implies \lim_{t \to +\infty} x(t, x_0) = x^*$$

The set $B = \{x_0 \in D : \lim_{t \to +\infty} x(t, x_0) = x^*\}$ is called the basin of attraction of x^*. The equilibrium is attractive if this set contains a ball with center x^* and radius $\eta > 0$, in other words if the basin of attraction is a neighborhood of x^*; if solutions tend to the equilibrium when $t \to -\infty$, the equilibrium is said to be *repulsive*:

– *asymptotically stable*, if it is stable and attractive;

– *globally attractive in* $D_0 \subset D$ if D_0 is the basin of attraction of x^*;

– *globally asymptotically stable (GAS) in* $D_0 \subset D$, if it is stable and globally attractive in D_0.

In order to distinguish asymptotic stability from global asymptotic stability, it is said that a point is *locally asymptotically stable* (LAS) when it is asymptotically stable without being (or when it is not possible to demonstrate it) *globally asymptotically stable* (GAS).

EXAMPLE A1.4.– For the logistic equation (see example A1.3), the equilibrium points are $x_1^* = 0$ and $x_2^* = K$. The first is unstable and the second is LAS. Its basin of attraction is $(0, +\infty)$.

EXERCISE A1.7.– More generally, show that for a scalar equation $\frac{dx}{dt} = f(x)$, where $x \in \mathcal{D} \subset \mathbb{R}$ and $x \in \mathcal{D} \mapsto f(x) \in \mathbb{R}$, an isolated equilibrium point x^* is attractive if and only if $f(x) > 0$ for $x^* - \eta < x < x^*$ and $f(x) < 0$ for $x^* < x < x^* + \eta$ for a certain $\eta > 0$. Deduce therefrom that if $f(x)$ cancels out and changes sign, from positive to negative, when x crosses the equilibrium value when it increases, then this is LAS. Demonstrate that if the function changes signs, from negative to positive, when x reaches the equilibrium value when increasing, then this is a repulsive equilibrium on both sides, and thus unstable. Show that if $f(x)$ cancels without changing sign, the equilibrium is semi-stable (attractive on one side and repulsive on the other side). This exercise shows that for scalar equations, being attractive and being LAS are equivalent, since attractive implies stable.

Being stable does not imply being asymptotically stable. The converse is true by definition, namely that asymptotically stable implies stable. Although the terminology employed seems to suggest that asymptotically stable is weaker than stable. As pointed out in exercise A1.7, for scalar equations, attractive implies stable, but this does not hold in a higher dimension as shown in the following example.

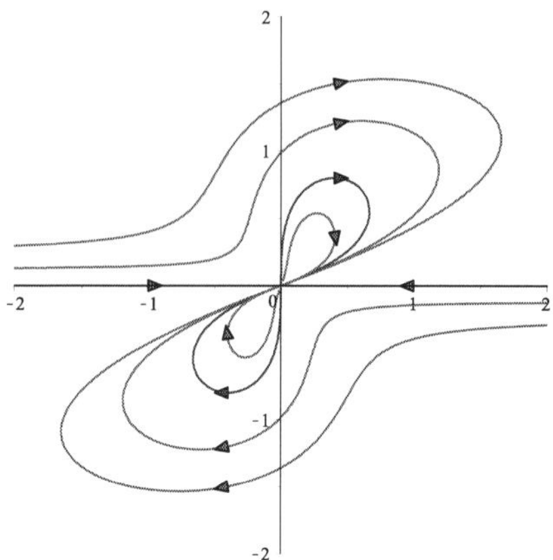

Figure A1.3. *Orbits [A1.12]. The two homocline orbits, in blue in the figure, form loops, filled with homocline orbits. For a color version of this figure, see www.iste.co.uk/harmand/chemostat.zip*

EXAMPLE A1.5 (Vinograd).– Consider the system:

$$\frac{dx}{dt} = \frac{x^2(y-x)+y^5}{(x^2+y^2)\left(1+(x^2+y^2)^2\right)}$$
$$\frac{dy}{dt} = \frac{y^2(y-2x)}{(x^2+y^2)\left(1+(x^2+y^2)^2\right)} \qquad \text{[A1.12]}$$

This system admits the origin as a single equilibrium point. Note that the straight line $y=0$ is invariant and that on this line, the system is written as:

$$\frac{dx}{dt} = \frac{-x}{1+x^4}$$

Therefore, $x=0$ is a globally attractive equilibrium point in the invariant axis $y=0$. It is more difficult to show that the origin is globally attractive for the full system [A1.12], but a numerical simulation (see Figure A1.3) shows that all solutions converge to the origin. The homocline orbits inside the two loops formed by the blue orbits correspond to solutions that start as close to the origin as we wish and that then move away to a considerable distance. As a result, the origin is not stable. The fact that the origin is not stable is due to these homocline orbits. For details and proofs of the existence of the two loops formed by homocline orbits, the reader can consult [HAH 67, p. 191].

We will see in section A1.2.2 that for linear systems, attractive implies stable, and that attractive, LAS and GAS are equivalent.

A1.2.2. *Stability of linear differentials systems*

A differential system is said to be linear if it is of the form:

$$\frac{dx}{dt} = Ax \qquad \text{[A1.13]}$$

where x is a vector of $\mathbb{R}^n$ and A a square matrix of order n:

$$x = \begin{bmatrix} x_1 \\ x_2 \\ \vdots \\ x_n \end{bmatrix}, \qquad A = \begin{bmatrix} a_{11} & a_{12} & \cdots & a_{1n} \\ a_{21} & a_{22} & \cdots & a_{2n} \\ \vdots & \vdots & \ddots & \vdots \\ a_{n1} & a_{n2} & \cdots & a_{nn} \end{bmatrix}$$

The behavior of these systems is very important for the study of any differential system (known as *nonlinear* as opposed to the linear case), of the form [A1.1], because a nonlinear differential system is approximated by its linearized form in the neighborhood of its singular points, see section A1.2.3.

We are first going to show how to integrate linear differential systems. To clarify the presentation, let us examine the case $n = 2$. The general case is similarly addressed (see the end of this section).

In order to solve this system of differential equations, a change of variables is employed that reduces it back to a simpler form. It is known [GOD 66] that for any square matrix, there is an invertible matrix T such that:

$$T^{-1}AT = J$$

where J is the Jordan form of matrix A. When $n = 2$, if the two eigenvalues of A are distinct, then J is a diagonal matrix with eigenvalues as diagonal elements. If there is a double eigenvalue, then a 1 may appear above the diagonal. Thereby, J is of one of the following three forms:

$$J = \begin{bmatrix} \lambda_1 & 0 \\ 0 & \lambda_2 \end{bmatrix}, \qquad J = \begin{bmatrix} \lambda & 0 \\ 0 & \lambda \end{bmatrix}, \qquad J = \begin{bmatrix} \lambda & 1 \\ 0 & \lambda \end{bmatrix} \qquad \text{[A1.14]}$$

where $\lambda_1 \neq \lambda_2$ (in the first case) or $\lambda = \lambda_1 = \lambda_2$ (in the second or the third case) are the eigenvalues of A.

The interest of this reduction of the matrix A to its Jordan form is that the change of variable $y = T^{-1}x$ transforms [A1.13] to:

$$\frac{dy}{dt} = Jy \qquad\qquad\qquad \text{[A1.15]}$$

Indeed, from $x = Ty$, it is deduced that $\frac{dx}{dt} = T\frac{dy}{dt}$. Thus:

$$T\frac{dy}{dt} = Ax = ATy$$

Consequently, $\frac{dy}{dt} = T^{-1}ATy = Jy$.

Since the Jordan form [A1.14] is very simple, equation [A1.15] can be integrated, almost as simply as the scalar equation [A1.3] had been integrated. By using $x = Ty$, is then deduced $x(t)$.

A1.2.2.1. *Node, saddle*

If the eigenvalues are real and distinct $\lambda_1 \neq \lambda_2$, then J is of the first form [A1.14]. As a result, [A1.15] is written as:

$$\frac{dy_1}{dt} = \lambda_1 y_1, \qquad \frac{dy_2}{dt} = \lambda_2 y_2$$

Therefrom, it is deduced that $y_1(t) = c_1 e^{\lambda_1 t}$ and $y_2(t) = c_2 e^{\lambda_2 t}$, where c_1 and c_2 are arbitrary constants. If we take $c_1 = 0$, we obtain as orbits the axis $y_1 = 0$, and if taking $c_2 = 0$, we obtain as orbits the axis $y_2 = 0$. The images of these axes by the linear mapping $x = Ty$ are the straight lines V_1 and V_2, see Figure A1.4. Note that V_1 and V_2 are the eigensubspaces of matrix A associated with the real and distinct eigenvalues λ_1 and λ_2.

Figure A1.4(a) illustrates the case of real and strictly negative eigenvalues, $\lambda_1 < \lambda_2 < 0$, showing that all the solutions tend to 0 when $t \to +\infty$. The origin is GAS: this is known as a *stable node*. When the eigenvalues are real and strictly positive, all solutions tend toward the origin when $t \to -\infty$. The origin is repulsive and therefore unstable: this is referred to as an *unstable node*.

Figure A1.4(b) illustrates the case of real eigenvalues $\lambda_1 < 0 < \lambda_2$, showing that the only solutions which tend to 0 when $t \to +\infty$, are those originating from initial conditions that belong to V_1 and that the only solutions which tend to 0 when $t \to -\infty$, are those that originate from initial conditions that belong to V_2. The *stable subspace* (or the *stable separatrix*) of the system is denoted by V_1 and V_2 designates the *unstable subspace* (or the *unstable separatrix*). The origin is unstable: this is known as a *saddle point.*

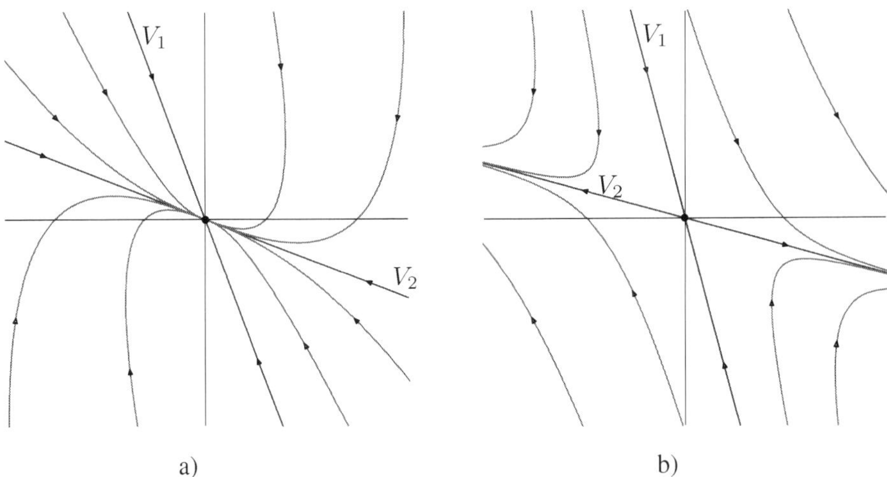

Figure A1.4. *Phase portraits: a) stable node $\lambda_1 < \lambda_2 < 0$ (the case of the unstable node $0 < \lambda_1 < \lambda_2$ is obtained by changing the orientation of the arrows on the orbits); b) saddle $\lambda_1 < 0 < \lambda_2$. For a color version of this figure, see www.iste.co.uk/harmand/chemostat.zip*

If the eigenvalues are equal $\lambda_1 = \lambda_2 = \lambda$, then, when A is diagonalizable, J is diagonal, of the second form [A1.14]. As a result, [A1.15] can be written as:

$$\frac{dy_1}{dt} = \lambda y_1, \qquad \frac{dy_2}{dt} = \lambda y_2$$

We find the same solutions as previously: $y_1(t) = c_1 e^{\lambda t}$ and $y_2(t) = c_2 e^{\lambda t}$ with c_1 and c_2 as arbitrary constants. The corresponding orbits are represented in Figure A1.5(b), in the case $\lambda < 0$. The origin is GAS: this is known as a *stable (degenerated) node.*

If A is not diagonalizable, then J is of the third form [A1.14] and [A1.15] is written as:

$$\frac{dy_1}{dt} = \lambda y_1 + y_2, \qquad \frac{dy_2}{dt} = \lambda y_2$$

EXERCISE A1.8.– Show that $y_1(t) = (c_1 + c_2 t)\, e^{\lambda t}$ and $y_2(t) = c_2 e^{\lambda t}$ where c_1 and c_2 are constants.

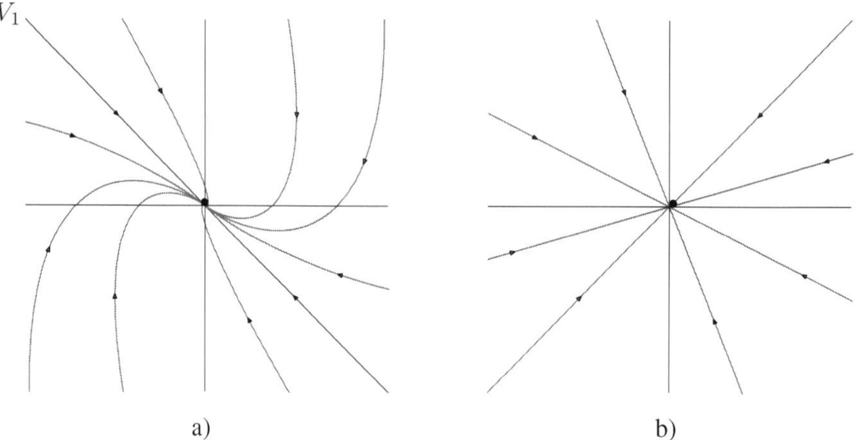

a)   b)

Figure A1.5. *Phase portraits of a degenerated stable node $\lambda_1 = \lambda_2 < 0$: a) non-diagonalizable case; b): diagonalizable case. The case of the unstable node $0 < \lambda_1 = \lambda_2 < 0$ is obtained by changing the orientation of the arrows on the orbits. For a color version of this figure, see www.iste.co.uk/harmand/chemostat.zip*

If we take $c_2 = 0$, we obtain as orbits the axis $y_2 = 0$. The image of this axis by the linear mapping $x = Ty$ is the straight line V_1 (Figure A1.5(a)). The eigensubspace of the matrix A is associated with the double eigenvalue λ. Figure A1.5(b) illustrates the case of real and strictly negative eigenvalues, $\lambda_1 = \lambda_2 < 0$, showing that all the solutions tend to 0 when $t \to +\infty$. The origin is GAS: this is known as a *stable (degenerated) node*. If $\lambda > 0$, then all solutions tend toward the origin when $t \to -\infty$. The origin is repulsive (therefore unstable): this is known as an *unstable (degenerated) node*.

A1.2.2.2. *Focus, center*

Since the matrix A has real coefficients, if its values λ_1 and λ_2 are complex, they are necessarily distinct and conjugated:

$$\lambda_1 = \alpha + i\omega, \qquad \lambda_2 = \alpha - i\omega$$

with $\omega > 0$. By taking linear combinations of the complex solutions $e^{\lambda_1 t}$ and $e^{\lambda_2 t}$, real solutions are found which are linear combinations of solutions of the form $e^{\alpha t}\cos(\omega t)$ and $e^{\alpha t}\sin(\omega t)$.

Figure A1.6(a) illustrates the case of eigenvalues with strictly negative real parts, $\alpha < 0$, showing that all the solutions tend to 0 when $t \to +\infty$. The origin is called a *stable focus*. When the eigenvalues have strictly positive real parts, all solutions tend toward the origin when $t \to -\infty$. The origin is repulsive (thus unstable): this is known as an *unstable focus*.

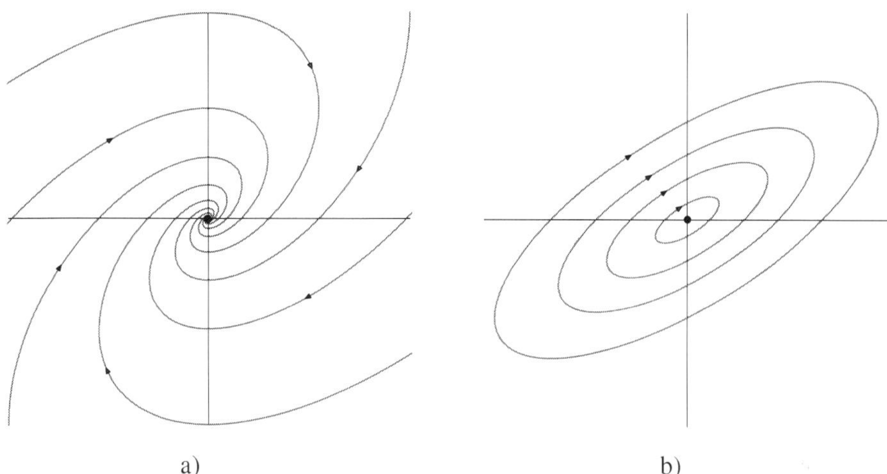

a)   b)

Figure A1.6. *Portrait phases: a) stable focus $\lambda_1 = \alpha + i\omega$, $\lambda_2 = \alpha - i\omega$, $\alpha < 0$ (the case of the unstable focus $0 < \alpha$ is obtained by changing the orientation of the arrows on the orbits); b) center $\alpha = 0$. For a color version of this figure, see www.iste.co.uk/harmand/chemostat.zip*

Figure A1.6(b) illustrates the case of two complex conjugate eigenvalues of real part equal to zero, $\alpha = 0$. All solutions are periodic. The origin is stable, but not LAS: it is said that the origin is a *center*.

The two constants c_1 and c_2 which appear in the solutions are determined according to the initial conditions $x_1(0)$ and $x_2(0)$. We thus recover the result of the existence and uniqueness of solutions when an initial condition is defined. The solutions are defined for all $t \in \mathbb{R}$. This result is true for all linear differential systems, not just those of dimension 2.

The detailed study done in dimension 2 has shown that solutions converge to 0 when t tends to infinity if and only if the two eigenvalues λ_1 and λ_2 have strictly negative real parts. This property is true in any dimension. We have the following result (where $\| \cdot \|$ designates a norm on $\mathbb{R}^n$) in which only the case known as *hyperbolic*, where all eigenvalues have non-zero real parts, is considered.

THEOREM A1.3.– Let $\lambda_1, \cdots, \lambda_n$ be the eigenvalues of A. Let $x(t, x_0)$ be the solution of [A1.13] such that $x(t, x_0) = x_0$:

– if there exists k such that $Re\lambda_k > 0$, then there exist initial conditions x_0 arbitrarily close to 0, such that $\lim\limits_{t\to+\infty} \|x(t, x_0)\| = +\infty$ (if $x_0 \neq 0$);

– if $Re\lambda_k < 0$ for $k = 1\cdots n$, then for every initial condition $x_0 \in \mathbb{R}^n$, $\lim\limits_{t\to+\infty} \|x(t, x_0)\| = 0$.

Based on theorem A1.3, the following result on the stability of the origin for a linear system can be deduced.

THEOREM A1.4.– For the system $\frac{dx}{dt} = Ax$, the origin is GAS in $\mathbb{R}^n$, if and only if all of the eigenvalues of matrix A have strictly negative real parts. If there exists an eigenvalue with a positive real part, then the origin is unstable.

Therefore, for linear systems, the terms attractive, LAS and GAS are equivalent.

A1.2.3. *Linearization*

The first step in the qualitative study of a differential system consists of the study of the system in the neighborhood of its equilibrium points. Consider the system [A1.11] and let x^* be an equilibrium point (therefore $f(x^*) = 0$). It is assumed that f admits continuous partial derivatives. The Taylor expansion of f in the neighborhood of x^* is written as:

$$f(x^* + y) = Ay + g(x), \text{ with } A = \frac{\partial f}{\partial x}(x^*) = \left[\frac{\partial f_i}{\partial x_j}(x^*)\right]$$

and $g(x)$ contains the higher order terms. The nonlinear system [A1.11] can thus be written as:

$$\frac{dx}{dt} = f(x) = Ay + g(x), \text{ with } y = x - x^* \tag{A1.16}$$

Since $\|g(x)\|$ is small compared to $\|x - x^*\|$, it is shown that, under certain conditions, which are specified in the following, the nonlinear system [A1.16] "behaves" as the following linear system:

$$\frac{dy}{dt} = Ay, \tag{A1.17}$$

called the *linearized system of the system at the point* x^*. We have the following result.

THEOREM A1.5.– If all of the eigenvalues of the matrix $A = \frac{\partial f}{\partial x}(x^*)$ have strictly negative real parts, then x^* is LAS for [A1.16]. Therefore, all orbits originating from a point quite close to x^* converge to x^* when $t \to +\infty$. It is said that x^* is a *well*.

If there is an eigenvalue of $A = \frac{\partial f}{\partial x}(x^*)$ having a strictly positive real part, then x^* is unstable for [A1.16]. If all of the eigenvalues of the matrix $A = \frac{\partial f}{\partial x}(x^*)$ have strictly positive real parts, then x^* is repulsive (and therefore unstable). The orbits of [A1.16] originate from x^*: it is said that x^* is a *source*.

This result shows that if the linearized system of the plan system is a node or a stable focus (respectively, unstable), then the equilibrium point of the nonlinear system is a well (respectively, a source). The nonlinear system therefore behaves locally as its linearized system. It will be seen in section A1.2.4 that when the linearized system is a saddle, then the nonlinear system behaves also like its linearized system.

DEFINITION A1.5.– *The equilibrium point x^* of [A1.16] is referred to as* locally exponentially stable (LES), *if there exist $\alpha > 0$, $\beta > 0$ and $\eta > 0$ such that for any initial condition x_0, we have*:

$$\|x_0 - x^*\| < \eta \implies \text{for all } t \geq 0, \|x(t, x_0) - x^*\| \leq \beta \|x_0 - x^*\| e^{-\alpha t}$$

It can be said that it is *globally exponentially stable in $D_0 \subset D$*, if there is $\alpha > 0$, $\beta > 0$ such that for any initial condition $x_0 \in D_0$ and for all $t \geq 0$, we have

$$\|x(t, x_0) - x^*\| \leq \beta \|x_0 - x^*\| e^{-\alpha t}$$

EXERCISE A1.9.– Show that if x^* is LES, then it is LAS. Show that the converse is false: to this end, consider the scalar equation $\frac{dx}{dt} = -x^3$ and show that the origin is LAS but not LES.

We demonstrate [HAL 69] that if f admits continuous partial derivatives, then x^* is LES if and only if all of the eigenvalues of the matrix $A = \frac{\partial f}{\partial x}(x^*)$ have negative real parts. theorem A1.5 exactly states that if all the eigenvalues of the linearized system have strictly negative real parts, then the equilibrium is LES and thus LAS, as it has been noticed in the previous exercise.

EXAMPLE A1.6.– This characterization of LES provides a means to find without performing computations that the origin of the scalar equation $\frac{dx}{dt} = -x^3$ is LAS (and even GAS) but not LES. As a matter of fact, the eigenvalue of the linearized system is equal to 0.

Since the eigenvalues λ_1 and λ_2 of a square matrix of dimension 2 are solutions to the equation:

$$\det(A - \lambda I) = \lambda^2 - \mathrm{Tr}A\lambda + \det A = 0$$

they have strictly negative real parts if and only if $\det A > 0$ and $\operatorname{Tr} A < 0$. Therefore, we have the following proposition.

PROPOSITION A1.6.– For a differential system of the plan $\frac{dx}{dt} = f(x)$, an equilibrium point x^* is LES, if and only if $\det A > 0$ and $\operatorname{Tr} A < 0$, where $A = \frac{\partial f}{\partial x}(x^*)$.

A1.2.4. *Stable manifold and unstable manifold*

We have seen in section A1.2.2 that in the plane, there exists two straight lines for a saddle point:

$$E^s = V_1, \qquad E^u = V_2$$

where V_1 and V_2 are the eigensubspaces associated with the eigenvalues $\lambda_1 < 0 < \lambda_2$, see Figure A1.4(b), such that:

$$x_0 \in E^s \Rightarrow x(t, x_0) \in E^s \text{ and } \lim_{t \to +\infty} x(t, x_0) = 0$$

$$x_0 \in E^u \Rightarrow x(t, x_0) \in E^u \text{ and } \lim_{t \to -\infty} x(t, x_0) = 0$$

where $x(t, x_0)$ is the solution of the linear differential system $\frac{dx}{dt} = Ax$ having initial condition x_0. Remember that the two vector subspaces of dimension 1, E^s and E^u, are called the stable and unstable subspaces of the origin. We have the following result that shows that if the linearization of a nonlinear system is a saddle, then the nonlinear system behaves locally around an equilibrium point as its linearized system at this equilibrium point. This is true in the sense that it also presents orbits tangent to subspaces E^s and E^u, which tend to x^* when $t \to +\infty$, for orbits tangent to E^s, and when $t \to -\infty$, for orbits tangent to E^u.

THEOREM A1.6.– Consider the differential system in the plane $\frac{dx}{dt} = f(x)$ defined in a domain $\mathcal{D}$ of $\mathbb{R}^2$. Let x^* be an equilibrium point, such that the eigenvalues of $A = \frac{\partial f}{\partial x}(x^*)$ are real and have opposite signs There exists a unique curve W^s, tangent to E^s at point x^*, called the *stable manifold x^** and a unique curve W^u, tangent to E^u at point x^*, called the *unstable manifold x^**, such that:

$$x_0 \in W^s \Rightarrow x(t, x_0) \in W^s \text{ and } \lim_{t \to +\infty} x(t, x_0) = x^*$$

$$x_0 \in W^u \Rightarrow x(t, x_0) \in W^u \text{ and } \lim_{t \to -\infty} x(t, x_0) = x^*$$

Therefore, the curve (stable manifold) W^s is invariant. It consists of three orbits: the point x^* itself and two particular orbits, which tend toward the saddle point when $t \to +\infty$. These are the only orbits (apart from the singular point itself) which verify this property. These latter are referred to as the *stable separatrices* of the saddle. Similarly, the curve (unstable manifold) W^s is invariant. It consists of three orbits: the point x^* itself and two special orbits, which tend toward the saddle point when $t \to -\infty$. These are the only orbits (apart from the singular point itself) which verify this property. These latter are referred to as the *unstable separatrices* of the saddle. See illustrations of W^s, W^u, E^s and E^u in Figures A1.7(b) and A1.8(b), in the case of system [A1.2].

The notion of stable and unstable subspaces can be extended to any dimension: E^s appears as the direct sum of all the generalized eigensubspaces corresponding to eigenvalues of negative real part and E^u as the direct sum of all the generalized eigensubspaces corresponding to eigenvalues of positive real part. Care should be taken in case of complex eigenvalues when defining E^s and E^u because in this case eigensubspaces are complex. We have expressed the theorem in the plane, but it holds for any dimension. It is rightly said that W^s and W^u are stable and unstable *manifolds* because in larger dimension, they are surfaces or objects of higher dimension referred to as manifolds. Consequently, for *hyperbolic* points (that is those whose eigenvalues of the linearized system have real nonzero parts), there exist a stable manifold W^s and an unstable manifold W^u that have the same dimension as E^s and E^u and which verify the properties set out in theorem A1.6. We do not provide details about the general case which goes beyond the scope that we have set out in this book. The reader who wishes to know more may consult the literature (for example [HIR 77]).

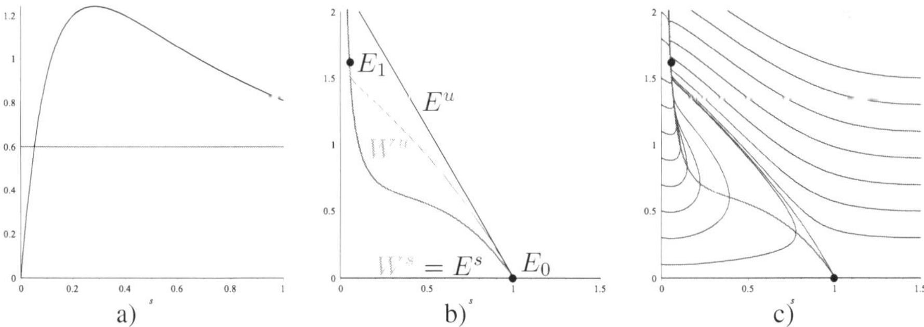

Figure A1.7. *Phase portrait of* [A1.2] *with* $q(s) = \frac{3s}{0.2+s+2.5s^2}$, $p(s) = \frac{q(s)}{1+0.5s}$, $D = 1$, $S_{in} = 1$ *and* $D_1 = 0.6$. *a) The graph of* $q(s)$. *b) The graph of* $\pi(s)$, *the separatrices* W^s *and* W^u *and sub-spaces* E^s *and* E^u *of* E_0. *c) The orbits, in blue, converge to* E_1. *For a color version of this figure, see www.iste.co.uk/harmand/chemostat.zip*

EXERCISE A1.10.– Consider system [A1.2]. Assume that $q(S_{in}) > D_1$. Show that E_0 is a saddle. Show that the stable manifold W^s of this saddle E_0 is equal to the stable eigensubspace E^s of the linearized system at this point:

$$W^s = E^s = \{(s,x) \in \mathcal{D} : s \geq 0, x = 0\}$$

Show that there is a positive solution $(s(t), x(t))$ such that:

$$\lim_{t \to -\infty} x(t) = 0 \text{ and } \lim_{t \to -\infty} s(t) = S_{in}$$

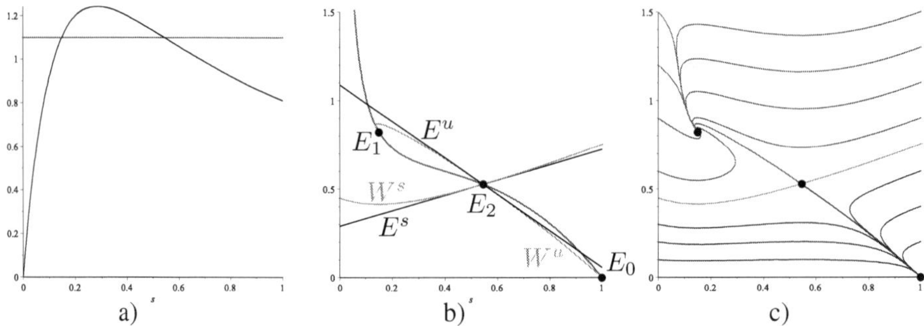

Figure A1.8. *Phase portrait of* [A1.2] *with the parameters considered in Figure A1.7, with the exception of* $D_1 = 1.1$. *a) The graph of* $q(s)$. *b) The graph of* $\pi(s)$, *the separatrices* W^s *and* W^u *and sub-spaces* E^s *and* E^u *of* E_0. *c) The stable manifold* W^s *of* E_2 *(in green) separates* $\mathcal{D}$ *into two areas which are the basins of attraction of* E_1 *and* E_0. *For a color version of this figure, see www.iste.co.uk/harmand/chemostat.zip*

The corresponding orbit is the separatrix W^u of E_0. This separatrix, as well as the unstable subspace E^u of the linearized system, is represented in Figure A1.7. This figure shows that, with the chosen parameter values, equation $q(s) = D_1$ admits a single solution s^* smaller than S_{in} (Figure A1.7(a)). Therefore, there is a single positive equilibrium $E_1 = (s^*, x^*) \approx (0.052, 1.621)$ which is LES because $q'(s^*) > 0$ and $\pi'(s^*) < 0$, see exercise A1.11. The equilibrium point $E_0 = (1, 0)$ is a saddle because $q(S_{in}) > D$. The graph of $\pi(s) = \frac{D(S_{in} - s)}{p(s)}$ is drawn in red (Figure A1.7(b)). The separatrices W^s and W^u of E_0, in green in the figure, are tangent to the stable and unstable vector subspaces, E^s and E^u, respectively, in black in the figure. Note that W^s and E^s are superposed on the axis $x = 0$ and that W^u is tangent to E^u at point E_0 and that it converges to E_1. All orbits, except those in $W^s = E^s$ converge to the positive equilibrium E_1 (Figure A1.7(c)).

EXERCISE A1.11.– Consider the system [A1.2]. Let $E_1 = (s^*, x^*)$ be a positive equilibrium, such that $q(s^*) = D_1$, $S_{in} > s^*$ and $x^* = \pi(s^*)$, see exercise A1.1.

Show that this equilibrium is LES for the system [A1.2] if and only if $q'(s^*) > 0$ and $\pi'(s^*) < 0$.

EXERCISE A1.12.– Consider the system [A1.2]. It is assumed that the equation $q(s) = D_1$ admits two solutions, and two only: s_1^* and s_2^*, such that $0 < s_1^* < s_2^* < S_{in}$, $q'(s_1^*) > 0$ and $q'(s_2^*) < 0$. Show that washout E_0 is a well. We denote by $E_1 = (s_1^*, x_1^*)$ and $E_2 = (s_2^*, x_2^*)$ the positive equilibria corresponding to s_1^* and s_2^*, with $x_1^* = \pi(s_1^*)$ and $x_2^* = \pi(s_2^*)$. Assume that $\pi'(s_1^*) < 0$. Show that only E_1 is a well and that E_2 is a saddle. The separatrices of the saddle E_2, as well as the one of the linearized system, are represented in Figure A1.8. This figure shows that, with the chosen parameter values, the equation $q(s) = D_1$ admits two solutions s_1^* and s_2^* smaller than S_{in} (Figure A1.8(a)). Consequently, there are two positive equilibria $E_1 = (s_1^*, x_1^*) \approx (0.147, 0.832)$ which are LES because $q'(s_1^*) > 0$ and $\pi'(s_1^*) < 0$ and $E_2 = (s_2^*, x_2^*) \approx (0.548, 0.527)$ is a saddle because $q'(s_2^*) < 0$, see exercise A1.11. The washout E_0 is LES because $q(S_{in}) < D$. The graph of $\pi(s) = \frac{D(S_{in}-s)}{p(s)}$ is drawn in red (see Figure A1.7(b)). The separatrices W^s and W^u of E_2, in green in the figure, are tangent to the stable and unstable vector subspaces, E^s and E^u respectively, in black in the figure. The stable manifold W^s of E_2 (in green in the figure) splits $\mathcal{D}$ into two areas which are the basins of attraction of E_1 and E_0: orbits drawn in red converge to E_1, those in blue converge to E_0 (Figure A1.8(c)).

A1.3. Limit sets

The solutions of a differential system can converge toward some points (which are then equilibrium points as shown in proposition A1.4). They can also converge toward more complicated sets, known as limit sets, and that we will study in this section.

A1.3.1. *Definitions and notations*

DEFINITION A1.6.– *The ω-limit set $\omega(x_0)$ of a point x_0 of $\mathcal{D}$ is the set of $y \in \mathcal{D}$ such that there exists a sequence $t_k \to +\infty$ such that $x(t_k, x_0) \to y$ when $k \to +\infty$. A point y is in the α-limit set $\alpha(x_0)$ if there exists a sequence $t_k \to -\infty$ such that $x(t_k, x_0) \to y$ when $k \to +\infty$.*

EXERCISE A1.13.– Show that if x_0 is an equilibrium point or a periodic point, then:

$$\omega(x_0) = \alpha(x_0) = \gamma(x_0)$$

Show that if a and b are in the same orbit, then $\omega(a) = \omega(b)$ and $\alpha(a) = \alpha(b)$, such that the limit sets of an orbit are defined as being limit sets of one of its points.

We have the following results [PER 13] which are very helpful for studying the asymptotic behavior of the solutions.

PROPOSITION A1.7.– A limit set is closed and invariant. If an orbit is positively bounded, then its ω-limit set is non-empty, compact and connected.

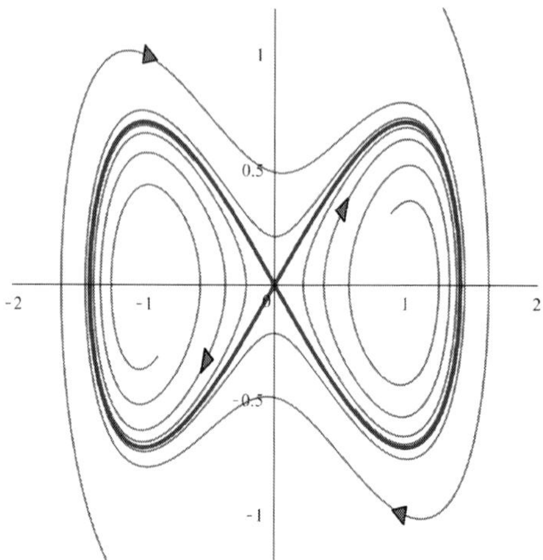

Figure A1.9. *Solutions of [A1.18], for the case $\mu = 1/2$. The two homocline orbits γ_1 and γ_2, in blue in the figure, form with the equilibrium point $E_0 = \{(0,0)\}$ the polycles $\{E_0\} \cup \gamma_1$, $\{E_0\} \cup \gamma_2$ and $\{E_0\} \cup \gamma_1 \cup \gamma_2$ which are the ω-limit sets of all orbits, with the exception of equilibrium points E_0, $E_1 = (1,0)$ and $E_2 = (-1,0)$ and orbits γ_1 and γ_2. For a color version of this figure, see www.iste.co.uk/harmand/chemostat.zip*

EXAMPLE A1.7.– Consider the system:

$$\begin{aligned} \frac{dx}{dt} &= y \\ \frac{dy}{dt} &= x - x^3 - \mu y \left(y^2 - x^2 + \frac{x^4}{2} \right) \end{aligned} \qquad [A1.18]$$

where μ is a real parameter. This system admits $E_0 = (0,0)$, $E_1 = (1,0)$ and $E_2 = (-1,0)$ as equilibrium points. Let:

$$H(x,y) = y^2 - x^2 + \frac{x^4}{2}$$

It follows that:

$$\frac{dH}{dt} = -4\mu y^2 H$$

Consequently, the set $H = 0$ is invariant. This set, which has the shape of a figure eight, is the reunion of three orbits: $\gamma_0 = \{E_0\}$ which is an equilibrium point, and the two homocline orbits of equations:

$$\gamma_1 = \left\{ (x, y) \in \mathbb{R}^2 : y = \pm x \sqrt{1 - \frac{x^2}{2}}, 0 < x \le \sqrt{2} \right\}$$

$$\gamma_2 = \left\{ (x, y) \in \mathbb{R}^2 : y = \pm x \sqrt{1 - \frac{x^2}{2}}, -\sqrt{2} \le x < 0 \right\}$$

that both admit $\{E_0\}$ as α-limit and ω-limit sets:

$$\alpha(\gamma_1) = \omega(\gamma_1) = \alpha(\gamma_2) = \omega(\gamma_2) = \{E_0\}$$

If the initial condition is chosen inside one of the loops of the figure eight, but different from E_1 and E_2, the solution converges toward this loop. If the initial condition is chosen outside the figure-eight curve, then the solution spirals to the figure-eight curve. In this case, the ω-limit set of an orbit different from equilibrium points or orbits γ_1 and γ_2 is one of the following three forms:

$$\gamma_0 \cup \gamma_1, \quad \gamma_0 \cup \gamma_2, \quad \gamma_0 \cup \gamma_1 \cup \gamma_2$$

Such limit sets are called polycycles. See Figure A1.9. The notion of polycyle will be described more precisely in section A1.3.3.

A1.3.2. *Butler–McGehee theorem*

THEOREM A1.7 (Butler–McGehee [SMI 95]).– If an omega-limit set ω contains a hyperbolic equilibrium point a, but it is not equal to $\{a\}$, then ω contains a point $b \neq a$ of the stable manifold $W^s(a)$, as well as a point $c \neq a$ of the unstable manifold $W^u(a)$.

This theorem will be used to show that the limit set of a trajectory of the chemostat model [A1.2] cannot contain neither the washout steady-state, when it is unstable (saddle point), nor any point of the axis $x = 0$.

PROPOSITION A1.8.– Consider the system [A1.2]. Let $(s(t), x(t))$ be a trajectory of initial condition (s_0, x_0) verifying $x_0 > 0$. If $q(S_{in}) > D_1$, the ω-limit set of this trajectory is non-empty and cannot contain any point of the axis $x = 0$.

Demonstration. According to proposition A1.7, the limit set is non-empty and compact. It will be denoted by ω. Note that ω cannot be equal to $\{E_0\}$. Indeed, as it has been seen in exercise A1.10, E_0 is a saddle whose stable manifold W^s is the

x-axis. Since E_0 is the ω-limit sets of points of W^s only and since $x_0 > 0$, the point (s_0, x_0) does not belong to W^s and therefore $\omega \neq \{E_0\}$.

Suppose then that $E_0 \in \omega$. As it has been showed that $\omega \neq \{E_0\}$, according to theorem A1.7 of Butler– McGehee, we thereof deduce that ω contains a point $(s_1, 0)$ of $W^s(E_0)$ such that $s_1 \neq S_{in}$. As ω is invariant, it contains the orbit $\gamma(s_1, 0)$. Two cases must be considered: $s_1 > S_{in}$ or $s_1 < S_{in}$. If $s_1 > S_{in}$, then the orbit:

$$\gamma(s_1, 0) = \{(s, 0) \in \mathcal{D} : s > S_{in}\}$$

is not bounded, thus it cannot be included in the compact set ω. If $s_1 < S_{in}$, the following trick is used: the functions $p(s)$ and $q(s)$ which are defined for $s \geq 0$ are extended to $s < 0$, by putting $p(s) = q(s) = 0$ for $s < 0$. As a result, the system [A1.2] is defined in the entire domain $\mathbb{R}^2$. The domain $\mathcal{D}$ is positively invariant and since the initial condition (s_0, x_0) is chosen in $\mathcal{D}$, $\omega \subset \mathcal{D}$. It follows that (for this new system defined in $\mathbb{R}^2$):

$$\gamma(s_1, 0) = \left\{(s, 0) \in \mathbb{R}^2 : s < S_{in}\right\}$$

which contains points where $s < 0$, therefore it cannot be included in ω because this limit set is included in $\mathcal{D}$.

A1.3.3. *Asymptotically autonomous systems*

The vector field f considered in [A1.1] does not depend on the variable t. It is then said that [A1.1] is an *autonomous* differential system. More generally, let us consider a *non-autonomous* differential system whose second member is also dependent on the variable t:

$$\frac{dx}{dt} = f(t, x) \tag{A1.19}$$

where $f : x \in (a, +\infty) \times \mathcal{D} \rightarrow f(x) \in \mathbb{R}^n$ is a continuous function, a a real number and $\mathcal{D}$ a subset of $\mathbb{R}^n$. Analogously to autonomous differential systems, one defines the concept of solution: a function $t \mapsto x(t)$ is a solution of [A1.19] if it is defined in an interval $I \subseteq (a, +\infty)$, if it takes its values in $\mathcal{D}$, if it is continuously differentiable on I and if, for any $t \in I$, we have $\frac{dx}{dt}(t) = f(t, x(t))$. It is shown that if f is sufficiently regular (for example if it admits continuous partial derivatives with respect to variables x_i, $i = 1 \cdots n$), then for all $(t_0, x_0) \in (a, +\infty) \times \mathcal{D}$, [A1.19] admits a unique solution $x(t)$ verifying $x(t_0) = x_0$. It is denoted by $x(t, t_0, x_0)$ to remind its dependence with respect to t_0 and x_0. If this solution is positively bounded, that is if it remains in a compact subset of $\mathcal{D}$ it is then defined for all $t \geq 0$ and its ω-limit set $\omega(t_0, x_0)$ is defined as in definition A1.6.

DEFINITION A1.7.– *A point y is inside the limit set $\omega(t_0, x_0)$ if and only if there exists a sequence $t_k \to +\infty$ such that $y = \lim_{k \to \infty} x(t_k, t_0, x_0)$.*

The purpose of this section is to study the limit sets of a particular class of non-autonomous system.

DEFINITION A1.8.– *The differential system* [A1.19] *is said to be* asymptotically autonomous *if there exists a continuous function $g : \mathcal{D} \to \mathbb{R}^n$ such that $f(t, x)$ converges to $g(x)$, where $t \to +\infty$, uniformly on any compact subset of $\mathcal{D}$.*

The differential system:

$$\frac{dx}{dt} = g(x) \tag{A1.20}$$

is called the *limit system* of [A1.19]. Markus [MAR 56] has shown the following result.

THEOREM A1.8 (Markus [MAR 56]).– The set ω-limit $\omega(t_0, x_0)$ of a positively bounded solution $x(t, t_0, x_0)$ of [A1.19] is non-empty, compact, connected and it is invariant according to [A1.20]. Moreover, if x^* is an LES equilibrium point of [A1.20], then there exists a neighborhood V of x^* and time T such that, if $x(t, t_0, x_0) \in V$ for a $t > T$, we then have $\omega(t_0, x_0) = \{x^*\}$.

In this theorem, it is necessary to assume that the solution being considered is positively bounded. Indeed, it could tend to infinity, such that its ω-limit set would be empty, as shown in the following example.

EXAMPLE A1.8.– Consider the system:

$$\begin{aligned}
\frac{dx}{dt} &= -x + x^2 y \\
\frac{dy}{dt} &= -y
\end{aligned} \tag{A1.21}$$

This system admits $E_0 = (0, 0)$ as unique equilibrium point. The second equation of [A1.21] admits as solutions the functions:

$$y(t) = y_0 e^{-t}$$

where y_0 is a constant. Therefore, the system [A1.21] is equivalent to the non-autonomous equation:

$$\frac{dx}{dt} = -x + x^2 y_0 e^{-t}$$

This equation is asymptotically autonomous and admits as limit system the equation:

$$\frac{dx}{dt} = -x$$

of which all solutions converge to 0. It is not possible to deduce therefrom that all the solutions of [A1.21] converge toward $(0,0)$. In fact, if we set:

$$H(x,y) = xy$$

it follows that:

$$\frac{dH}{dt} = H(H-2)$$

Consequently, the set $H = 2$ is invariant. This set is a hyperbole that splits the plane into two regions: region $xy < 2$, which is the basin of attraction of E_0 and region $xy \geq 2$ which corresponds to non-positively bounded solutions.

Assuming that the equilibrium points of [A1.20] are isolated, and that any solution of [A1.20] converges toward one of these equilibrium points, does any positively bounded solution of [A1.19] also converge toward an equilibrium of [A1.20]? Without additional assumptions about the equilibrium points of [A1.20], the answer is no as it is shown in the following example.

EXAMPLE A1.9 (Thieme).– Consider the system:

$$
\begin{aligned}
\frac{dx_1}{dt} &= (1-r)x_1 - (\beta|x_2| + x_3^2)x_2 \\
\frac{dx_2}{dt} &= (1-r)x_2 + (\beta|x_2| + x_3^2)x_1 \\
\frac{dx_3}{dt} &= -\gamma x_3
\end{aligned}
\qquad\text{[A1.22]}
$$

where $r = \sqrt{x_1^2 + x_2^2}$, and $\beta > 0$, $\gamma > 0$ are real parameters. Since $x_3(t) = x_3(0)e^{-\gamma t}$, the system [A1.22] is asymptotically autonomous and admits as limit system:

$$
\begin{aligned}
\frac{dx_1}{dt} &= (1-r)x_1 - \beta|x_2|x_2 \\
\frac{dx_2}{dt} &= (1-r)x_2 - \beta|x_2|x_1
\end{aligned}
\qquad\text{[A1.23]}
$$

The limit system has a three equilibrium points: $E_0 = (0,0)$, $E_1 = (1,0)$ and $E_2 = (1,0)$. The axis $x_2 = 0$ is invariant. Similarly, the circle $r = 1$ is invariant: it is the reunion of four orbits, the equilibrium points E_1 and E_2 as well as of the orbits:

$$\gamma_1 = \{r = 1, x_2 > 0\}, \quad \gamma_2 = \{r = 1, x_2 < 0\}$$

Any solution with initial condition $x_2(0) > 0$ tends toward E_2 and any solution with initial condition $x_2(0) < 0$ tends toward E_1, see Figure A1.10. It is proved [THI 92, THI 94] that if $\beta > 2\gamma$, then the ω-limit set of any solution of [A1.22] such that $x_3(0) \neq 0$ and $r(0) \neq 0$ is the circle $r = 1$. This circle is called a *polycycle*, because:

$$\omega(\gamma_2) = E_1 = \alpha(\gamma_1), \qquad \omega(\gamma_1) = E_2 = \alpha(\gamma_2)$$

Note that the circle is not an ω-limit set of [A1.23].

EXERCISE A1.14.– Show the properties of [A1.23] set out in the previous example.

Thieme [THI 92] has shown that if we exclude the possibility of having polycycles in the limit system [A1.20], then the answer is yes to the question previously raised. More specifically, we refer to a *polycycle* of a differential system as a finite set of equilibrium points E_1, ..., E_k, and orbits γ_1, ..., γ_k, such that:

$$\omega(\gamma_k) = E_1 = \alpha(\gamma_1) \text{ and } \omega(\gamma_i) = E_{i+1} = \alpha(\gamma_{i+1}), \quad i = 1 \cdots k - 1$$

THEOREM A1.9 (Thieme).– If the equilibrium points of [A1.20] are isolated, if any solution of [A1.20] converges toward one of the equilibrium points, and if [A1.20] does not admit any polycycle, then any solution of [A1.19] also converges toward an equilibrium point of [A1.20].

There is a still more general result than theorem A1.9 and which is expressed based on the concept of chain recurrent set, but that we will not expose in this book because it would lead to too abstract expansions. The reader who wishes to know more is invited to consult [MIS 95].

A1.3.4. *Asymptotic behavior in the plane*

THEOREM A1.10 (Poincaré–Bendixon).– If an orbit γ of a differential system of the plane is positively bounded and if $\omega(\gamma)$ does not contain any equilibrium points, then $\omega(\gamma)$ is a periodic orbit.

Here follows an application.

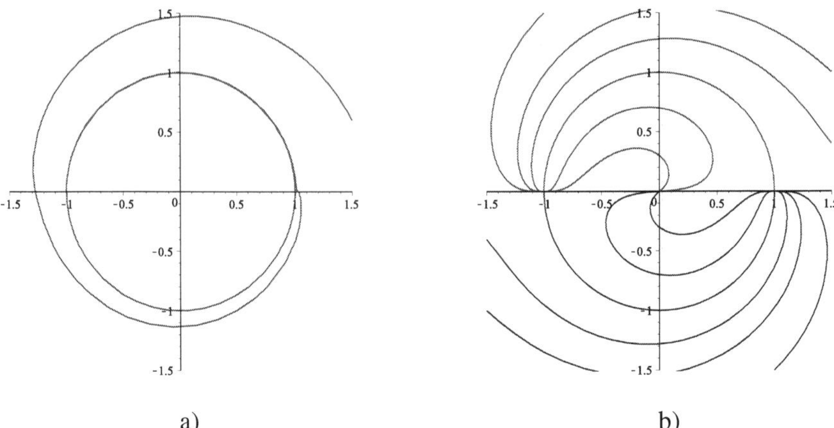

a)   b)

Figure A1.10. *System [A1.22], with $\beta = 3$, $\gamma = 1$. a) Projection in the plane (x_1, x_2) of the solution of initial condition $(1.5, 0.6, 2)$ showing the convergence toward the circle $r = 1$. b) The phase portrait of the limit system [A1.23] showing that all the solutions of initial condition $x_{20} > 0$, in red in the figure, converge to $E_2 = (-1, 0)$ and that all the solutions of initial condition $x_{20} < 0$, in blue in the figure, converge to $E_1 = (1, 0)$. For a color version of this figure, see www.iste.co.uk/harmand/chemostat.zip*

EXERCISE A1.15.– Let us consider the system [A1.2] in which $s \mapsto q(s)$ is strictly increasing. Assume that $q(S_{in}) > D_1$. Let s^* be the unique solution of the equation $q(s) = D_1$ and let $E_1 = (s^*, x^*)$ be the corresponding positive equilibrium. with $x^* = \pi(s^*)$, where $\pi(s) = \frac{D(S_{in} - s)}{p(s)}$. Assume that $\pi'(s_1^*) > 0$. Show that E_1 is a source and that it is surrounded by a periodic orbit. This limit cycle, whose existence is guaranteed by the Poincaré–Bendixon theorem, is illustrated in Figure A1.11. This figure shows the three orbits originating from points $(1, 7)$, $(0, 1)$ and (s_0, x_0) very close to $E_1 = (s^*, x^*) \approx (0.3, 6.532)$. They converge toward a cycle that surrounds E_1 which is a source because $\pi'(s^*) > 0$.

The following proposition provides a criterion that allows us to ensure that a differential system of the plane does not have a periodic orbit.

THEOREM A1.11 (Dulac–Bendixon criterion).– Consider the system $\frac{dx}{dt} = f(x)$ in the plane. If $\operatorname{div} f = \frac{\partial f_1}{\partial x_1} + \frac{\partial f_2}{\partial x_2}$ does not cancel in a region Ω of the plane, then Ω does not contain any periodic orbit.

Here follows an application of that criterion that shows how it can be used, as well as the Poincaré–Bendixon theorem to demonstrate that an LES equilibrium is GAS.

PROPOSITION A1.9.– Let us consider the system [A1.2] where $s \mapsto p(s)$ and $s \mapsto q(s)$ are strictly increasing functions for $s > 0$ and zero at 0. If $q(S_{in}) > D_1$, then E_1 is GAS in the positive quadrant $\mathcal{D}_0 = \{(s, x) \in \mathcal{D} : x > 0\}$.

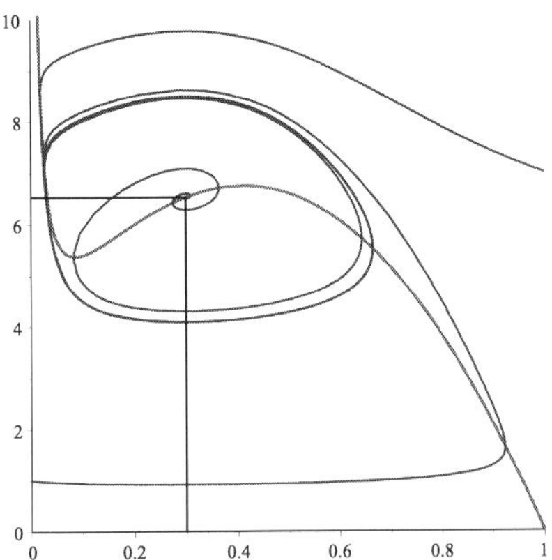

Figure A1.11. *Limit cycle of [A1.2] with $D = 1$, $D_1 = 0.75$, $S_{in} = 1$,
$q(s) = \frac{s}{0.1+s}$, $p(s) = \frac{q(s)}{1+20s}$. The graph of $\pi(s)$ is drawn in red. For a
color version of this figure, see www.iste.co.uk/harmand/chemostat.zip*

Demonstration. To show that the equilibrium E_1 is GAS as soon as it exists, the change
of variable $y = \ln(x)$ is used which is defined in $\mathcal{D}_0$ because $x > 0$. The system [A1.2]
is written as:

$$
\begin{aligned}
\frac{ds}{dt} &= D(S_{in} - s) - p(s)e^y \\
\frac{dy}{dt} &= q(s) - D_1
\end{aligned}
\qquad [A1.24]
$$

The divergence of the vector field $f(s, y) = \begin{bmatrix} D(S_{in} - s) - p(s)e^y, \\ q(s) - D_1, \end{bmatrix}$ is:

$$
\mathrm{div} f(s, y) = -D - e^y p'(s)
$$

Since $p'(s) > 0$, the divergence is negative and according the Bendixon–Dulac
criterion, the system cannot have a periodic solution.

Let (s_0, x_0) an original condition such that $x_0 > 0$. According to exercise A1.4,
the solutions are positively bounded. Therefore, the limit set ω of orbit $\gamma(s_0, x_0)$ is
non-empty and included in $\mathcal{D}$. Following proposition A1.8, this limit set cannot
contain a point of the axis $x = 0$. It is thus included in $\mathcal{D}_0$. If it does not contain the

point E_1 then, according to the Poincaré–Bendixon theorem, solutions converge toward a cycle, but the system does not present any due to the divergence property. As a result, it contains the point E_1. It is equal to it because E_1 is LES. This shows that E_1 is GAS in $\mathcal{D}_0$.

The following is another application of this criterion, which shows that the positive equilibrium of the density-dependent model:

$$\begin{aligned}
\frac{ds}{dt} &= D(S_{in} - s) - \mu(s, x)x \\
\frac{dx}{dt} &= (\mu(s, x) - D_1)x
\end{aligned} \qquad \text{[A1.25]}$$

considered in Chapter 2 is GAS, as soon as it exists. Recall (see section 2.3.3 of Chapter 2) that, if we assume that the function μ is increasing in s and decreasing in x then, if $\mu(S_{in}, 0) > D_1$, the system [A1.25] admits a unique positive equilibrium. This equilibrium is LES, see proposition 2.3 of Chapter 2. This equilibrium is given by $E_1 = (s_c, x_c)$ where s^* is the unique solution of the equation:

$$\frac{D}{D_1}(S_{in} - s) = \psi(s)$$

and $x_c = \psi(s_c)$, in which $s \mapsto \psi(s)$ is the characteristic at equilibrium, see definition 2.2 of Chapter 2. The equilibrium E_1 is in reality GAS, as stated in the following result.

PROPOSITION A1.10.– Let us consider the system [A1.25] in which $\mu(0, x) = 0$, $\frac{\partial \mu}{\partial s} > 0$ and $\frac{\partial \mu}{\partial s} < 0$. If $\mu(S_{in}, 0) > D_1$ then $E_1 = (s_c, x_c)$ is GAS in the positive quadrant $\mathcal{D}_0 = \{(s, x) \in \mathcal{D} : x > 0\}$.

Demonstration. In order to show that the equilibrium E_1 is GAS as soon as it exists, the change of variables $y = \ln(x)$ is used which is defined in $\mathcal{D}_0$ because $x > 0$. The system [A1.25] is written as:

$$\begin{aligned}
\frac{ds}{dt} &= D(S_{in} - s) - \mu\left(s, e^y\right)e^y \\
\frac{dy}{dt} &= \mu\left(s, e^y\right) - D_1
\end{aligned} \qquad \text{[A1.26]}$$

The divergence of the vector field $f(s, y) = \begin{bmatrix} D(S_{in} - s) - \mu\left(s, e^y\right)e^y, \\ \mu\left(s, e^y\right) - D_1, \end{bmatrix}$ is:

$$\operatorname{div} f(s, y) = -D - e^y \frac{\partial \mu}{\partial s}\left(s, e^y\right) + e^y \frac{\partial \mu}{\partial x}\left(s, e^y\right)$$

Since $\frac{\partial \mu}{\partial s} > 0$ and $\frac{\partial \mu}{\partial s} < 0$, the divergence is negative and according the Bendixon–Dulac criterion, the system cannot have a periodic solution.

Let (s_0, x_0) an initial condition such that $x_0 > 0$. As in exercise A1.4, it is shown that the solutions are positively bounded. Consequently, the limit set ω of the orbit $\gamma(s_0, x_0)$ is non-empty and included in $\mathcal{D}$. Similarly to the proof of proposition A1.8, by using the Butler–McGehee theorem, we show that this limit set cannot contain a point of the axis $x = 0$. It is thus included in $\mathcal{D}_0$. If it does not contain the point E_1 then, according to the Poincaré–Bendixon theorem, solutions converge toward a cycle, but the system does not present any due to the divergence property. As a result, it contains the point E_1. It is equal thereto because E_1 is LES. This shows that E_1 is GAS in $\mathcal{D}_0$.

A1.4. Supplements

A1.4.1. *Lyapunov functions*

This amounts to addressing once more the general case of dimension n, considering an autonomous system:

$$\frac{dx}{dt} = f(x) \qquad\qquad [A1.27]$$

defined on a domain D of $\mathbb{R}^n$. It is assumed that f admits continuous partial derivatives. Let $V : U \longrightarrow \mathbb{R}$, with $U \subset D$, be a function that also admits continuous partial derivatives. Let us denote by $\dot{V}(x)$ the scalar product of the gradient:

$$\frac{\partial V}{\partial x} = \left(\frac{\partial V}{\partial x_1}, \cdots , \frac{\partial V}{\partial x_n} \right)$$

of V and of the vector field f:

$$\dot{V}(x) = \sum_{i=1}^{n} \frac{\partial V}{\partial x_i}(x) f_i(x)$$

$\dot{V}(x)$ is called the derivative of the function V in the direction of the vector field f. This derivative is also referred to as the Lie derivative of V along f and is denoted by $L_f V$. This terminology is justified by the following formula. Let $x(t, x_0)$ be the solution of [A1.27] of initial condition x_0. According to the derivation theorem of composite functions, it follows that:

$$\frac{d}{dt}V\left(x\left(t,x_0\right)\right) = \sum_{i=1}^{n} \frac{\partial V}{\partial x_i}\left(x\left(t,x_0\right)\right)\frac{dx_i}{dt}(t) = \sum_{i=1}^{n} \frac{\partial V}{\partial x_i}\left(x\left(t,x_0\right)\right) f_i\left(x\left(t,x_0\right)\right)$$

Consequently, $\frac{d}{dt}V\left(x\left(t,x_0\right)\right) = \dot{V}\left(x\left(t,x_0\right)\right)$ and by setting $t = 0$, the following formula is thereof derived:

$$\dot{V}(x_0) = \frac{d}{dt}V\left(x\left(t,x_0\right)\right)_{|t=0}$$

We have the following result (for the demonstration, see [HAL 69, HIR 74]):

THEOREM A1.12.– Let $x^* \in D$ be an equilibrium point of [A1.27] and let V be a function defined on a neighborhood U of x^*, admitting continuous partial derivatives and such that:

a) $V(x^*) = 0$ and $V(x) > 0$ if $x \neq x^*$;

b) $\dot{V}(x) \leq 0$ if $x \neq x^*$;

then x^* is stable. If in addition, we have:

c) $\dot{V}(x) < 0$ if $x \neq x^*$, then x^* is LAS.

If in addition, we have:

d) $\|V(x)\|$ tends to $+\infty$ when x tends toward the boundary of U, then x^* is GAS in U.

A function V verifying (a) and (b) is called a Lyapunov function. If in addition (c) is satisfied, then V is said to be a strict Lyapunov function. If in addition (d) is satisfied, then V is said to be a proper Lyapunov function.

This theorem is widely used to prove the global stability of an equilibrium point. However, the difficulty in applying this theorem is that there is no general method to construct a Lyapunov function. Moreover, it is even harder to find strict Lyapunov functions. We have the following result, known as the Krasovskii–LaSalle principle of invariance, which allows for asymptotic stability results to be obtained when the Lyapunov function is not strict.

THEOREM A1.13.– Let $x^* \in D$ be an equilibrium point of [A1.27] and let V be a Lyapunov function defined in the neighborhood U of x^*. Let M be the largest invariant subset included in the set:

$$\mathcal{V} = \left\{ x \in D : \dot{V}(x) = 0 \right\}$$

Then, the ω-limit set of any orbit originating from a point of U is included in M.

The following is an application of this result. Let us consider the model [A1.2] that we write again here as:

$$\frac{ds}{dt} = D(S_{in} - s) - p(s)x$$
$$\frac{dx}{dt} = (q(s) - D_1)x \tag{A1.28}$$

where $s \mapsto p(s)$ and $s \mapsto q(s)$ are positive functions for $s > 0$ and equal to zero at 0. It should be recalled that a positive equilibrium of the form $E_1 = (s^*, x^*)$ where s^* is defined by $q(s^*) = D_1$ and $x^* = \pi(s^*)$ with:

$$\pi(s) = \frac{D(S_{in} - s)}{p(s)} \tag{A1.29}$$

This equilibrium exists if and only if $S_{in} > s^*$. As it has been seen in exercise A1.11, this equilibrium is LES if and only if $q'(s^*) > 0$ and $\pi'(s^*) < 0$. Is it GAS? Example 2.39 of Chapter 2, and the simulations presented in Figure 2.19, show that this condition is not sufficient to ensure that the equilibrium be GAS. Indeed, in Figures 2.19(b) and 2.19(c), an LES equilibrium point surrounded by two cycles can be seen, one unstable and the other stable. Therefore, it is not GAS. The following theorem provides a sufficient condition for which the equilibrium point is GAS. Note that this condition, which means that the line $x = x^*$ does only meet the isocline $x = \pi(s)$ at the point x^*, is not a necessary condition for the equilibrium point to be GAS, as shown by the simulations represented in Figures 2.18(c) and 2.18(d).

THEOREM A1.14.– Suppose that the equation $q(s) = D_1$ admits a single solution $s = s^*$ such that $s^* < S_{in}$ and that the equation $\pi(s) = \pi(s^*)$ admits a single solution in the interval $(0, S_{in})$. Therefore, the equilibrium point $E_1 = (s^*, x^*)$ is GAS in the positive quadrant for the system [A1.28], that is for any solution $(s(t), x(t))$ having initial condition $s(0) \geq 0$, $x(0) > 0$, we have $\lim_{t \to +\infty} s(t) = s^*$ and $\lim_{t \to +\infty} x(t) = x^*$.

Demonstration. It can be assumed that $s(0) < S_{in}$. Consider the Lyapunov function:

$$V(s, x) = \int_{s^*}^{s} \frac{q(\sigma) - D_x}{p(\sigma)} d\sigma + \int_{x^*}^{x} \frac{\xi - x^*}{\xi} d\xi \tag{A1.30}$$

This function is positive and only cancels out in E_1. The derivative of V along the trajectories is:

$$\dot{V}(s, x) = \frac{\partial V}{\partial s}\frac{ds}{dt} + \frac{\partial V}{\partial x}\frac{dx}{dt} = \frac{q(s) - D_1}{p(s)}\frac{ds}{dt} + \frac{x - x^*}{x}\frac{dx}{dt}$$

Using [A1.28] and $x^* = \pi(s^*)$ yields:

$$\dot{V}(s,x) = \frac{q(s) - D_1}{p(s)} \left[D(S_{in} - s) - p(s)x \right] + \left[x - \pi(s^*) \right] \left[q(s) - D_1 \right]$$

After simplification of terms $x[q(s) - D_1]$ and using [A1.29], we obtain:

$$\dot{V}(s,x) = [q(s) - D_x] \left[\pi(s) - \pi(s^*) \right]$$

Since the equation $q(s) = D_1$ admits a single solution $s = s^*$ such that $s^* < S_{in}$ and the equation $\pi(s) = \pi(s^*)$ admits a single solution in the interval $(0, S_i n)$, for all $s \in (0, S_{in})$, we have:

$$(q(s) - D_1)(\pi(s) - \pi(s^*)) < 0, \text{ for } s \neq s^* \tag{A1.31}$$

According to [A1.31], $\dot{V}(s,x)$ is negative for $0 < s < S_{in}$ and cancels if and only if $s = s^*$. Consequently:

$$\mathcal{V} = \{\dot{V}(s,x)\} = \{s = s^*\}$$

The largest invariant set contained in $\mathcal{V}$ is constituted of the equilibrium $\{E_1\}$. Using therorem A1.13, it can be deduced that solutions tend to E_1.

Unlike the arguments based on the Dulac criterion and the Poincaré–Bendixon theorem used in section A1.3.4 to demonstrate that an equilibrium point is GAS, and which are valid in the plane only, the proof techniques based on Lyapunov functions encompass higher dimensions, see section A1.5.

EXERCISE A1.16.– Demonstrate theorem A1.14 using the Lyapunov function:

$$V = (s,x)x^* \int_{s^*}^{s} \frac{q(\sigma) - D_1}{D(S_{in} - \sigma)} d\sigma + \int_{x^*}^{x} \frac{\xi - x^*}{\xi} d\xi \tag{A1.32}$$

A1.4.2. *Isoclines, trajectories traps*

The analysis of isoclines and the construction of trajectories traps are simple and effective methods for achieving the qualitative study of the solutions of a differential equation. *Unfortunately*, contrary to the Lyapunov functions method, it is usable only in dimension 2.

Through secondary school, one learns to qualitatively draw the graph of a function of f of $\mathbb{R}$ in itself. It is reminded that the principle thereof consists of establishing the monotony table:

x	$-\infty$			$+\infty$
f'	$-$	$+$	$-$	$-$
f	$\searrow$	$\nearrow$	$\searrow$	$\searrow$

$$[A1.33]$$

To this end, the values of x that cancel the derivative are determined, which defines *intervals* in which f is *monotonic*; then minima, maxima and behaviors at infinity are determined which makes it possible to draw the graph "by hand". For a differential system in dimension 2, a similar procedure can be utilized to analyze the "variations of trajectories" and trace them "by hand". This is known as the *method of isoclines*.

A vector field (see section A1.1.1):

$$(x,y) \mapsto \begin{bmatrix} f(x,y) \\ g(x,y) \end{bmatrix}$$

defined on a domain D of $\mathbb{R}^2$. must be viewed as vectors of origin (x,y) and end $(x+f(x,y), y+g(x,y))$ since $(f(x,y), g(x,y)$ is the velocity in (x,y) of the solution of:

$$\begin{aligned} \frac{dx}{dt} &= f(x,y) \\ \frac{dy}{dt} &= g(x,y) \end{aligned}$$

$$[A1.34]$$

which goes through this point at a given time. The set of the trajectories that are parameterized curves: $t \mapsto (s(t), x(t))$ of the domain of the plane where is defined the vector field is called the *flow* associated with the differential system[1] (Figure A1.12).

Obviously, it is not possible to draw all of the trajectories of [A1.34] (there is an infinity of them) but we can try to draw the most significant ones: the equilibria indicating their type (saddle, focus and node), periodic solutions, the separatrices of the saddles and a few specific solutions. This is what is called the *phase portrait*.

---

1 This terminology can be explained in the following way: the water that runs along a river is the *flow*; the flow can be complex due to the presence of obstacles. The flow has a certain velocity at each point of the space: this is the vector field; the movement of a floater that floats will materialize a trajectory of the flow associated with this vector field.

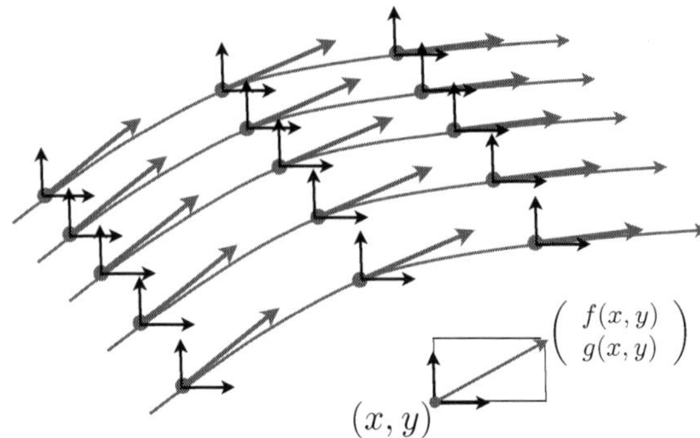

Figure A1.12. Vector field *and associated* flow. *For a color version of this figure, see www.iste.co.uk/harmand/chemostat.zip*

When considering an explicit system, a number of trajectories can be simulated with a computer to provide a better description of the phase portrait, but before addressing simulations, it is important to understand as many *phase portrait* properties as possible.

To illustrate the method of isoclines, we will use the Lotka–Volterra competition system which is the system:

$$\frac{dx}{dt} = \alpha(1 - ax - by)x$$
$$\frac{dy}{dt} = \beta(1 - cx - dy)y$$

[A1.35]

where all constants are positive.

DEFINITION A1.9.– *Isoclines. We call* isocline of x *of the system* [A1.34] *and denote by $\frac{dx}{dt} = 0$, the set of points where $f(x,y) = 0$. Similarly, the* isocline of y: $\frac{dy}{dt} = 0$ *is the set of points where $g(x,y) = 0$.*

REMARK A1.1.– Strictly speaking, "isocline" means "to have the same slope" and "so-called isocline of x" should be called "isocline" of zero slope or still "zero-cline" and "isocline of y", "isocline of infinite slope". The designations "isocline of x" and "isocline of y" are preferred which are more meaningful.

The equilibria of [A1.34] are the intersection points of the isoclines.

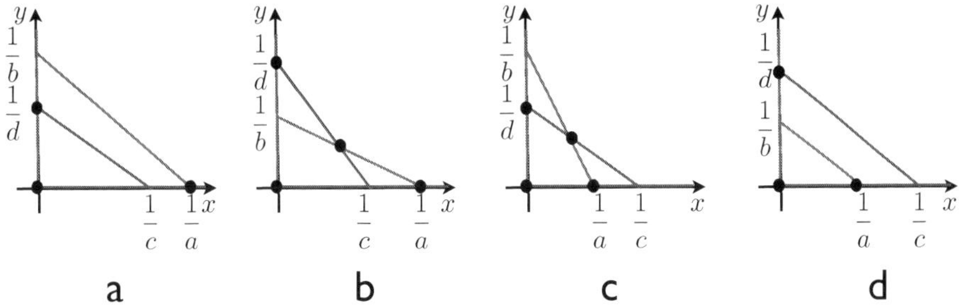

Figure A1.13. *Isoclines and Lotka–Volterra equilibria. For a color version of this figure, see www.iste.co.uk/harmand/chemostat.zip*

In Figure A1.13, the isocline of x is represented in red: $\frac{dx}{dt} = 0$ and in blue the isocline of the y: $\frac{dy}{dt} = 0$ of the Lotka–Volterra competition system [A1.35]. Each isocline consists of a line of negative slope and one of the axes. There are four possible cases if excluding non-generic cases where both lines are superposed. The representation with different colors is useful to immediately determine equilibrium points: these are the intersection points of the *red and blue*. There is a subtle difference between the cases (b) and (c): in case (c), the four equilibria form a convex set, which is not the case in (b). When the lines intersect (cases (b) and (c)), the isoclines determine *four* regions, *three* when they do not intersect. In each region, the signs of $\frac{dx}{dt}$ and $\frac{dy}{dt}$ are constant.

What has just been said about the particular Lotka–Volterra system will be true for any system [A1.34]. *Isoclines define a certain number of regions (generally, a finite number) of the phase space where the signs of $\frac{dx}{dt}$ and $\frac{dy}{dt}$ are constant*. In such a domain, the solutions of [A1.34] are *monotonic*.

PROPOSITION A1.11.– Let $\mathcal{D}$ be a domain where the system [A1.34] is monotonic. Any solution of [A1.34] originating from a point of $\mathcal{D}$:

– either leaves $\mathcal{D}$;

– or tends to $\hat{A} \pm \infty$;

– or still tends toward an equilibrium that belongs to $\mathcal{D}$ or to its boundary.

Demonstration. Let $t \mapsto (x(t), y(t))$ be a solution issued from a point of $\mathcal{D}$. We assume that it does not leave $\mathcal{D}$ and does not tend to infinity. Then, $t \mapsto x(t)$ is monotonic – because we are not leaving $\mathcal{D}$ – and bounded – since it does not tend to infinity. A monotonic bounded function tends toward a limit, therefore $x(t)$ tends toward a limit x_e and $y(t)$ tends toward a limit y_e. According to proposition A1.4, the point (x_e, y_e) is an equilibrium that belongs to $\mathcal{D}$ or to its boundary.

A1.4.3. *Application to Lotka–Volterra*

In Figure A1.14, we have represented the case (c) of Figure A1.13 of the isoclines of the Lotka–Volterra system. The equilibria are E_o, E_1, E_2, E_3. The figure shows how four domains have thus been delimited, I (turquoise), II (pink), III (yellow) and IV (purple) *where the system [A1.35] is monotonic* in agreement with the table of signs:

$$[A1.36]$$

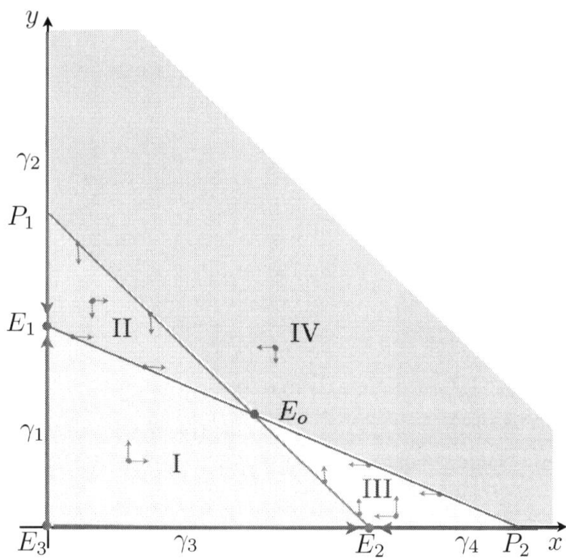

Figure A1.14. *Lotka–Volterra case no. 3. For a color version of this figure, see www.iste.co.uk/harmand/chemostat.zip*

It should be noted that domain IV is not bounded. The trajectories of [A1.35] for which $x = 0$ have been designated by γ_1, γ_2 and those for which $y = 0$ by γ_3, γ_4.

1) Let a trajectory of [A1.35] originate from a point inside II:

- it cannot leave along the segment $[P_1\ E_0]$ since at a point of this segment the vector:

$$\begin{bmatrix} f(x,y) \\ g(x,y) \end{bmatrix}$$

is vertical pointing downward, thus is strictly entering in II;

- for the same reason if cannot leave along $[E_1\ E_0]$;

- it cannot neither leave along γ_2 nor through E_1 or through E_0 since these are trajectories;

- since the trajectory cannot leave, it tends toward an equilibrium contained in II; this can only be E_0 because in II $t \mapsto x(t)$ is strictly increasing.

2) In a similar manner, a trajectory originating from a point inside of III tends toward the equilibrium E_0.

3) Let now a trajectory originate from a point inside of I:

- either it leaves I and then this can only be according to $[E_1\ E_0]$ or $[E_2\ E_0]$; it then enters domain II or III and then it tends to E_0 according to (1) or (2);

- or it stays inside I and then it tends to an equilibrium located in I. This cannot be E_1 because $t \mapsto x(t)$ is strictly increasing in I, nor through E_2 because $t \mapsto y(t)$ is strictly increasing in I, it is therefore E_o.

4) A trajectory issued from a point of IV can only "descend" and "go to the left"; it can directly tend toward E_0 or penetrate into II or III where it then tends toward E_0.

We have showed that for any initial condition of $(\mathbb{R}^{*+})^2$, the corresponding solution tends toward E_0. We have therefore proved that E_0 is attractive for all solutions originating from a point of $(\mathbb{R}^{*+})^2$. It will be verified that case (c) which we have just considered corresponds to the condition $ad > bc$ and that this implies that the Jacobian matrix in E_0 is stable. Therefore, E_0 is stable and consequently GAS.

EXERCISE A1.17.– Consider the three configurations (a), (b) and (d) of the isoclines of [A1.35] described in Figure A1.13. Show that:

– case (a) The equilibrium $(\frac{1}{a}, 0)$ is globally asymptotically stable;

– case (b) The equilibria $(\frac{1}{a}, 0)$ and $(0, \frac{1}{d})$ are stable;

– case (d) The equilibrium $(0, \frac{1}{d})$ is globally asymptotically stable.

A1.5. Bibliographic notes

The model [A1.2] has been considered by Arino, Pilyugin and Wolkowicz [ARI 03] as an extension of the conventional case of the chemostat model for which the function $y(s) = Y$ is constant and does not depend on the substrate density.

The theorem A1.14 has been obtained by Arino, Pilyugin and Wolkowicz (see [ARI 03], theorem 2.11) using the Lyapunov function [A1.32], and by Pilyugin and Waltman (see [PIL 03], lemma 2.3) using the Lyapunov function [A.30].

The Lyapunov functions have been used by various authors to prove the theorem of competitive exclusion in the chemostat model with different dilution rates. [HSU 78] can be cited for growth functions of the "Monod type" [WOL 92] for more general growth functions and [SAR 11, SAR 13] for yield rates depending on the substrate density. For more information on results in this area, the reader can consult [DEL 03, HSU 05, SAR 13]. Example 2.39 of Chapter 2 that presents an LES equilibrium surrounded by two limit cycles (Figure 2.19) has been proposed by [PIL 03].

Appendix 2

Indications for the Exercises

A2.1. Chapter 2 exercises

A2.1.1. *Exercise 2.1*

The solution $x(t)$ is obviously increasing since its derivative is positive. Because e_2 is an equilibrium, based on the uniqueness of the solutions it follows that $x(t) < e_2$, therefore that $x(t)$ is bounded, and then that $x(t)$ tends toward a limit $l \leq e_2$ when $t \to +\infty$. Assume that $l < e_2$; then there exists T such that for $t \geq T$ on $x(t) > f(l)/2$ and since:

$$x(t) = x(T) + \int_T^t f(x(s))ds \geq x(T) + (t - T)f(l)/2$$

that would imply that $x(t)$ would tend toward $+\infty$ with t which is contradictory. Analogously, it would be possible to see that $x(t)$ tends to e_1 when t tends to $-\infty$.

A2.1.2. *Exercise 2.2*

In Figure A2.1(a), the domain $\mathcal{D}_\varepsilon$ (in blue) is defined by the polygon ABCD and it will easily be verified that on its contour the vector field defined by [2.2] effectively points inward which shows that no trajectory can leave. Let $\varepsilon > 0$ be given and $\eta = \varepsilon/2$. Figure A2.1(b) shows that for any initial condition (s_o, x_o) such that $|s_o - S_{in}| \leq \eta$ and $|x_o \leq \eta$, the corresponding solution is such that $|s(t) - S_{in}| \leq \varepsilon$ and $|x(t) \leq \varepsilon$ which is the definition of stability (see section A1.2).

A2.1.3. *Exercise 2.3*

System [2.2] becomes:

$$\begin{aligned}
\frac{dz}{dt} &= D(S_{in} - z) \\
\frac{dx}{dt} &= \big(\mu(z - x) - D\big)x
\end{aligned} \qquad\qquad [A2.1]$$

whose Jacobian matrix at equilibrium is:

$$\begin{bmatrix}
-D & 0 \\
\mu'(S_{in} - x^*)x^* & -\mu'(S_{in} - x^*)x^*
\end{bmatrix} \qquad\qquad [A2.2]$$

In this triangular matrix, it can immediately be seen that stability is linked to the sign of $\mu'(S_{in} - x^*) = \mu'(s^*)$.

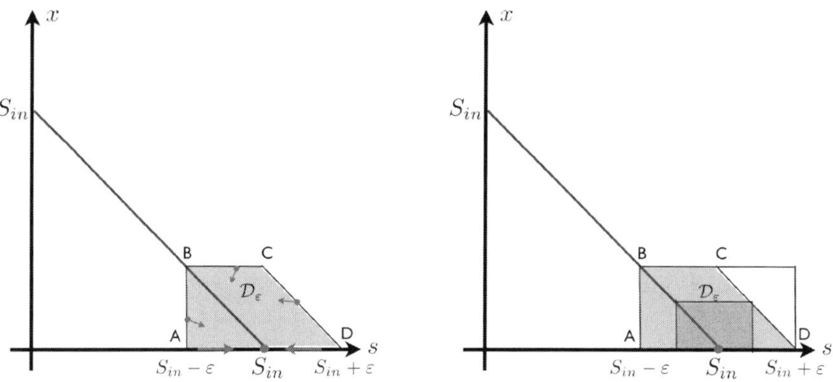

Figure A2.1. *Stability of* $(S_{in}, 0)$ *when* $D = \mu(S_{in})$. *For a color version of this figure, see www.iste.co.uk/harmand/chemostat.zip*

A2.2. Chapter 3 exercises

A2.2.1. *Exercise 3.1*

In the coordinates $(m, x_1, \cdots, x_n)$, the system is written as:

$$\begin{aligned}
\dot{m} &= D(S_{in} - m) \\
\dot{x}_i &= \mu_i\Big(m - \sum_{j=1}^{n} x_j\Big)x_i - Dx_i \quad i = 1 \cdots n
\end{aligned}$$

In E_0, the Jacobian is diagonal:

$$J(E_0) = \begin{bmatrix} -D & & & \\ & \mu_1(S_{in}) - D & & 0 \\ & & \ddots & \\ & 0 & & \mu_n(S_{in}) - D \end{bmatrix}$$

and the eigenvalues are present in the diagonal, as expected. In E_1, the Jacobian matrix is written block-wise:

$$J(E_1) = \left[\begin{array}{cc|ccc} -D & 0 & 0 & \cdots & 0 \\ \mu_1'(s_1^\star)x_1^\star & -\mu_1'(s_1^\star)x_1^\star & -\mu_1'(s_1^\star)x_1^\star & \cdots & -\mu_1'(s_1^\star)x_1^\star \\ \hline & & \mu_2(s_1^\star) - D & & 0 \\ & 0 & & \ddots & \\ & & 0 & & \mu_n(s_1^\star) - D \end{array} \right]$$

Since the block on the top left is triangular, its eigenvalues can be read in its diagonal: $-D$ and $-\mu_1'(s_1^\star)x_1^\star$; the other eigenvalues being given by the diagonal of the block (diagonal) on the bottom right.

A2.2.2. *Exercise 3.2*

We consider, without loss of generality, that the numbering is such that:

$$\lambda_1(D_1) < \lambda_2(D_2) < \cdots < \lambda_n(D_n).$$

Therefore, the only possible equilibria are the washout E_0 and equilibria E_i such that $s_i^\star = \lambda_i(D_i)$, $x_i^\star = S_{in} - s_i^\star$ and $x_j^\star = 0$ for $j \neq i$, as soon as $\lambda_i(D_i) < S_{in}$. The Jacobian matrix in E_0 is written as:

$$J(E_0) = \left[\begin{array}{c|ccc} -D & -\mu_1(S_{in}) & \cdots & -\mu_n(S_{in}) \\ \hline 0 & \mu_1(S_{in}) - D_1 & & 0 \\ \vdots & & \ddots & \\ 0 & 0 & & \mu_n(S_{in}) - D_n \end{array} \right]$$

and in E_1 (when $\lambda_1(D_1) < S_{in}$) :

$$J(E_1) = \left[\begin{array}{cc|ccc} -D - \mu_1'(s_1^\star)x_1^\star & -D & -\mu_n(s_1^\star) & \cdots & -\mu_n(s_1^\star) \\ \mu_1'(s_1^\star)x_1^\star & 0 & 0 & \cdots & 0 \\ \hline & & \mu_2(s_1^\star) - D_2 & & 0 \\ & 0 & & \ddots & \\ & & 0 & & \mu_n(s_1^\star) - D_n \end{array} \right]$$

We thus obtain the same statement as proposition 3.1 when replacing the break-even concentration $\lambda_i(D)$ by $\lambda_i(D_i)$.

A2.2.3. *Exercise 3.3*

1) By defining:

$$m = s + \sum_{i=1}^{n} x_i$$

the set Ω corresponds to the $(s, x) \in \mathbb{R}_+^{n+1}$ such that $m \leq S_{in}$ and we get:

$$\dot{m} = D(S_{in} - m)$$

For example, the variable m cannot cross $m = S_{in}$.

2) For $s > \lambda_1$ and $x_1 > 0$, we have:

$$\dot{x}_1 = (\mu_1(s) - D)x_1 > (\mu_1(\lambda_1) - D)x_1 = 0$$

On the other hand, when $x_1(0) > 0$, $x_1(t)$ remains positive for any t. Thus, $x_1(t)$ is strictly increasing for all t such that $s(t) > \lambda_1$.

3) For a solution that remains in C with $x_1(0) > 0$, $x_1(t)$ is an increasing function upper bounded by $x_1^\star = S_{in} - \lambda_1$, which thus admits a limit $\bar{x}_1 > 0$. Because $\dot{x}_1$ is uniformly continuous, $\dot{x}_1(t)$ converges to 0, and then $s(t)$ converges to λ_1. As a result, $q(t) = \sum_{i=2}^{n} x_i(t)$ converges to $\bar{q} = S_{in} - \lambda_1 - \bar{x}_1 \geq 0$. Since $s(t)$ converges to λ_1, which verifies $\mu_i(\lambda_1) < D$ for all $i \neq 1$, there exists $\epsilon > 0$ and $T > 0$ such that $\mu_i(s(t)) < D - \epsilon$ for all $t > T$. Now, it yields:

$$\dot{q}(t) = \sum_{i=2}^{n} (\mu_i(s(t)) - D)x_i(t) \leq -\epsilon q, \quad t > T$$

Therefore, $q(t)$ converges to 0 and thus $\bar{x}_1 = x_1^\star = S_{in} - \lambda_1$.

4) A solution that remains in $\mathcal{B}$ verifies:

$$\dot{x}_i = (\mu_i(s) - D)x_i \leq (\mu_i(\lambda_1) - D)x_i$$

for every i, that is:

$$x_i(t) \leq x_i(0)e^{(\mu_i(\lambda_1)-D)t}$$

As the numbers $\mu_i(\lambda_1) - D$ are negative for every $i \neq 1$, we can deduce that $x_i(t)$ tends to 0 for all $i \neq 1$. However, it follows that $\dot{x}_1(t) \leq 0$ for all t. Then, $x_1(t)$ is a monotonic decreasing function lower bounded by 0 that thus admits a limit $\bar{x}_1 \geq 0$. Furthermore, $s(t)$ also admits a limit equal to $\bar{s} = S_{in} - \bar{x}_1$, and that verifies $\bar{s} \leq \lambda_1$. It can be derived that $\bar{x}_1 \geq x_1^\star = S_{in} - \lambda_1 > 0$. Since $\dot{x}_1(t)$ is uniformly continuous, $\dot{x}_1(t)$ tends to 0, which implies $\bar{s} = \lambda_1$ and therefore $\bar{x}_1 = x_1^\star$.

5) In $s = \lambda_1$, x_1 necessarily verifies $x_1 < S_{in} - \lambda_1$ and we obtain:

$$\begin{aligned}
\dot{s} &= D(S_{in} - \lambda_1) - \sum_{i=1}^{n} \mu_i(\lambda_1)x_i \leq D(S_{in} - \lambda_1) - \mu_1(\lambda_1)x_1 \\
&= D(S_{in} - \lambda_1 - x_1) < 0
\end{aligned}$$

As a result, s cannot cross the boundary $s = \lambda_1$ from $\mathcal{B}$.

6) Let an initial condition be in Ω with $x_1(0) > 0$. Either the solution remains in $\mathcal{C}$ for any time and then converges to E_1, or it enters $\mathcal{B}$ and no longer leaves it. It then also converges toward E_1.

A2.2.4. *Exercise 3.4*

1) The derivative of μ is written as:

$$\mu'(s) = \frac{\mu_0(K_S - s^2/K_I)}{(K_S + s + s^2/K_I)^2}$$

Thus, μ is increasing on $[0, \sqrt{K_S K_I}[$ then decreasing on $]\sqrt{K_S K_I}, +\infty[$. Consequently, $\mu(s)$ reaches its maximum for $s = \sqrt{K_S K_I}$.

2) The solutions of $\mu(s) = D$ are solutions of the second- degree polynomial:

$$\frac{s^2}{K_I} + s\left(1 - \frac{\mu_0}{D}\right) + K_S = 0$$

whose discriminant is:

$$\Delta = \left(1 - \frac{\mu_0}{D}\right)^2 - 4\frac{K_S}{K_I}$$

When $D < \mu_0/(1 + 2\sqrt{K_S/K_I})$, Δ is positive and the two roots are positive and given by the expressions [3.12].

A2.2.5. *Exercise 3.5*

When the growth functions are monotonic, we have $\Lambda_i(D) =]\lambda_i(D), +\infty[$. Therefore, the set $\mathcal{E}(S_{in}, D)$ is written as:

$$\mathcal{E}(S_{in}, D) =] \min_i \lambda_i(D), +\infty[\cap]0, S_{in}[$$

If $\min_i \lambda_i(D) > S_{in}$, $\mathcal{E}(S_{in}, D)$ is empty and E_0 is the only equilibrium (locally exponentially) stable. If $\min_i \lambda_i(D) = \lambda_{i\star}(D) < S_{in}$, then $\mathcal{E}(S_{in}, D) =]\lambda_{i\star}(D), S_{in}[$ and $E_{i\star}$ is the only positive equilibrium, which is (locally exponentially) stable.

A2.2.6. *Exercise 3.6*

1) The equality:

$$\frac{\mu_{max}s}{K + s} = \frac{\mu_0 s}{K_S + s + s^2/K_I}$$

is verified for $s = 0$ or s solution of:

$$\mu_{max}(K_S + s + s^2/K_I) = \mu_0(K + s)$$

which admits two roots at most, being a second-degree polynomial.

2) The solutions of $\mu_1(s) = \mu_2(s)$ are 0, $1/3$ and 2. Furthermore, the graphs of these two functions have two intersections outside of 0.

3) Since it can be noted that we have $\max_s \mu_1(s) = 1/3 > \max_s \mu_2(s) = 3/10$, we obtain the operating diagram given in Figure A2.2, with the same color convention as for Figure 3.11.

A2.2.7. *Exercise 3.7*

We determine the sets:

$$\Lambda_1(D) =]17/30 - \sqrt{109}/30, 17/30 + \sqrt{109}/30[$$

$$\Lambda_2(D) =]5/2 - \sqrt{5}/2, 5/2 + \sqrt{5}/2[$$

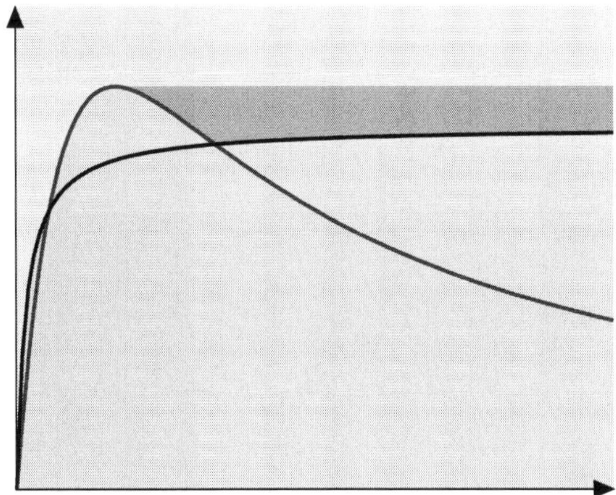

Figure A2.2. *Operating diagram in the plane (S_{in}, D) for exercise 3.6. For a color version of this figure, see www.iste.co.uk/harmandchemostat.zip*

It is observed that we have:

$$\bar{\lambda}_1(D) \;=\; 17/30 + \sqrt{109}/30 < 17/30 + \sqrt{1219}/30$$
$$=\; 28/30 < 1 < \lambda_2(D) = 5/2 - \sqrt{5}/2$$

and $\bar{\lambda}_2(D) < S_{in} = 5$. Therefore, this corresponds to case 5 of Table 3.2.

A2.3. Chapter 4 exercises

A2.3.1. *Exercise 4.1*

The Jacobian matrix in $(x_1^{\dagger}, x_2^{\dagger})$ is given by (we have omitted the argument of the functions):

$$\begin{bmatrix} -\partial_s\mu_1 + \partial_1\mu_1 & -\partial_s\mu_1 + \partial_2\mu_1 \\ -\partial_s\mu_2 + \partial_1\mu_2 & -\partial_s\mu_2 + \partial_2\mu_2 \end{bmatrix} \qquad\qquad \text{[A2.3]}$$

The trace of this matrix is strictly negative, the eigenvalues will have negative real parts if and only if the determinant is positive, namely:

$$\left(-\partial_s\mu_1 + \partial_1\mu_1\right)\left(-\partial_s\mu_2 + \partial_2\mu_2\right) > \left(-\partial_s\mu_1 + \partial_2\mu_1\right)\left(-\partial_s\mu_2 + \partial_1\mu_2\right) \quad \text{[A2.4]}$$

which is exactly the relation of proposition 4.1 (therefore, in this proposition, we can replace asymptotically by exponentially, which is slightly stronger).

A2.3.2. *Exercise 4.2*

We want to prove that, in the generic case, the coexistence equilibrium of the system [4.17] that we rewrite here is locally exponentially stable:

$$
\begin{cases}
\dfrac{ds}{dt} & = & D(S_{in} - s) - \sum_{i=1}^{N} \mu_i(s, x_i) x_i \\
\dfrac{dx_i}{dt} & = & (\mu_i(s, x_i) - D) x_i \qquad i = 1, \cdots, n
\end{cases}
$$

Let the equilibrium be:

$$
(s^*, \psi_1(s^*), \cdots, \psi_i(s^*), \cdots, \psi_n(s^*))
$$

The elements $a_{ij} \; : \; i, j = 0, 1 \cdots, N$ of the Jacobian matrix are:

$$
\begin{aligned}
a_{0,0} & = & -\left(D + \sum_{i=1}^{n} \partial_s \partial \mu_i(s^*, \psi_i(s^*)) \psi_i(s^*) \right) \\
a_{0,i} & = & -\partial_{x_i} \partial \mu_i(s^*, \psi_i(s^*)) \psi_i(s^*) + \mu_i(s^*, \psi_i(s^*)) \\
a_{i,0} & = & \partial_s \partial \mu_i(s^*, \psi_i(s^*)) \psi_i(s^*) \\
a_{i,i} & = & -\partial_{x_i} \partial \mu_i(s^*, \psi_i(s^*)) \psi_i(s^*) + \mu_i(s^*, \psi_i(s^*)) - D \\
a_{i,j} & = & 0 \qquad i \neq j
\end{aligned}
$$

Among the $\psi_i(s^*)$, some are equal to zero; the equations can be renumbered such that:

$$
i = p + 1, \cdots, n \iff \psi_i(s^*) = 0
$$

For $i = 1, \cdots, p$, we define:

$$
\begin{aligned}
a_i & = & \partial_s \mu(s^*, \psi_i(s^*)) \psi_i(s^*) \\
b_i & = & -\partial_{x_i} \partial \mu_i(s^*, \psi_i(s^*)) \psi_i(s^*) + \mu_i(s^*, \psi_i(s^*)) - D \\
& = & -\partial_{x_i} \partial \mu_i(s^*, \psi_i(s^*)) \psi_i(s^*)
\end{aligned}
$$

(because $\psi_i(s^*) > 0$ we have $\mu_i(s^*, \psi_i(s^*)) = D$) and for $i = p + 1 \cdots, n$:

$$
d_i = \mu_i(s^*, 0) - D
$$

We have $a_i > 0$, $b_i > 0$ and as we have assumed $s^* \neq \lambda_i$ we also have $d_i > 0$.

With these notations, the Jacobian matrix is written in blockwise triangular form:

$$J = \left(\begin{array}{c|c} A & B \\ \hline 0 & C \end{array} \right) \tag{A2.5}$$

where matrices A, B and C are, respectively:

$$A = \begin{pmatrix} -D - \sum_{i=1}^{p} a_i & b_1 - D & b_2 - D & \cdot & b_i - D & \cdot & b_p - D \\ a_1 & -b_1 & 0 & \cdot & \cdot & \cdot & 0 \\ \cdot & & & \cdot & \cdot & \cdot & 0 \\ a_i & 0 & \cdot & -b_i & 0 & \cdot & 0 \\ \cdot & & & \cdot & \cdot & \cdot & 0 \\ a_p & 0 & \cdot & \cdot & 0 & \cdot & -b_p \end{pmatrix} \tag{A2.6}$$

$$B = \begin{pmatrix} d_{p+1} + D & \cdot & \cdot & \cdot & d_i + D & \cdot & \cdot & \cdot & d_n + D \\ 0 & 0 & 0 & 0 & 0 & 0 & 0 & 0 & 0 \\ 0 & & & & & & & & 0 \\ 0 & 0 & 0 & 0 & 0 & 0 & 0 & 0 & 0 \end{pmatrix} \tag{A2.7}$$

$$C = \begin{pmatrix} -d_{p+1} & 0 & \cdot & \cdot & \cdot & \cdot & 0 \\ & \cdot & \cdot & \cdot & \cdot & \cdot & \\ 0 & \cdot & \cdot & -d_j & \cdot & \cdot & 0 \\ \cdot & \cdot & \cdot & & \cdot & \cdot & \\ 0 & \cdot & \cdot & & \cdot & \cdot & -d_n \end{pmatrix} \tag{A2.8}$$

The eigenvalues of matrix J are those of A and the diagonal matrix C which are all strictly negative real numbers.

To show the asymptotic stability of [4.17], we must show that the eigenvalues of A all have strictly negative real parts. To this end, we return to the definition of an eigenvector and its corresponding eigenvalue. Let thus $V = (v_0, v_1, \cdots, v_i, \cdots, v_p)^t$ be such that $JV = \lambda V$. From line no. 0 of [A2.6], we can deduce that:

$$\left(D + \sum_{i=1}^{p} a_i \right) v_0 + \sum_{i=1}^{p} (b_i - D) v_i = \lambda v_0 \tag{A2.9}$$

and from the nth line of [A2.6]:

$$a_i v_0 - b_i v_i = \lambda v_i$$

– either $v_0 = 0$, then since at least one of the v_i must be non-zero for this index, it follows that $\lambda = -b_i$ thus strictly negative;

– or $v_0 \neq 0$, then v_i can be expressed according to v_0, inserted in equation [A2.9] and simplified by v_0 which gives:

$$-\left(D + \sum_{i=1}^{p} a_i\right) = \lambda - \sum_{i=1}^{p}(b_i - D)\frac{a_i}{\lambda + b_i} \qquad [A2.10]$$

We want to show that the number $\lambda = \alpha + i\beta$ (eventually real if $\beta = 0$) such that $\alpha < 0$. Let us assume the contrary, namely that $\alpha \geq 0$. In [A2.10], the left-side member is a real which must be equal to the real part of the right-hand side. Let us denote by $\mathcal{R}(z)$ the real part of a complex z:

$$\mathcal{R}\left(\lambda - \sum_{i=1}^{p}(b_i - D)\frac{a_i}{\lambda + b_i}\right) = \alpha + \mathcal{R}\left(\sum_{i=1}^{p}(b_i - D)\frac{a_i}{\alpha + i\beta + b_i}\right) \quad [A2.11]$$

Since:

$$\mathcal{R}\left((b_i - D)\frac{a_i}{\alpha + i\beta + b_i}\right) = a_i\frac{(b_D)(\alpha + b_i)}{(\alpha + b_i)^2 + \beta^2} \leq a_i \qquad [A2.12]$$

from which we can deduce that:

$$\mathcal{R}\left(\lambda - \sum_{i=1}^{p}(b_i - D)\frac{a_i}{\lambda + b_i}\right) \geq \alpha - \sum_{i=1}^{p}a_i \geq -\sum_{i=1}^{p}a_i \qquad [A2.13]$$

However, according to [A2.10] and [A2.11], we have:

$$\mathcal{R}\left(\lambda - \sum_{i=1}^{p}(b_i - D)\frac{a_i}{\lambda + b_i}\right) = -\sum_{i=1}^{p}a_i - D < -\sum_{i=1}^{p}a_i \qquad [A2.14]$$

which is incompatible with [A2.13].

A2.3.3. *Exercise 4.3*

Point (1) is an immediate consequence of the notations. For point (2), let us calculate the Jacobian matrix J_i^* in $(S_{in}, 0, \cdots, x_{i_o}^*, 0, \cdots, 0)$. It is the matrix:

$$J_{i_o}^* = \begin{pmatrix} -D & 0 & 0 & \cdots & \cdots & 0 \\ 0 & A_1 & 0 & \cdots & \cdots & 0 \\ 0 & \cdots & \searrow & \cdots & \cdots & \cdot \\ B_{i_o} & B_{i_o} & \cdots & B_{i_o} & \cdots & B_{i_o} \\ 0 & 0 & \cdots & 0 & \searrow & 0 \\ 0 & \cdots & \cdots & \cdots & 0 & A_n \end{pmatrix} \qquad [A2.15]$$

with:

$$A_i = \nu_i(S_{in} - x_{i_o})\frac{1}{1 + x_{i_o}} - D \qquad B_{i_o} = (\nu_{i_o}(z - x)\frac{1}{1 + x} - D)'_{x=x_{i_o}} x_{i_o} \, [\text{A2.16}]$$

whose eigenvalues are A_i, $i \neq i_o$ and B_{i_o} which is negative; all the A_i with a rank lower than i_o are positive, the other negative thus only the matrix J_1^* is stable.

A2.4. Chapter 5 exercises

A2.4.1. *Exercise 5.1*

1) When it is assumed that only flocs of two individuals do exist, the specific attachment term depends only on the free biomass and not on the attached biomass. Furthermore, the simplest expression for the function α is $\alpha(u, v) = au$.

2) With $\beta(v) = b$, we get $g(x, p) = -ap^2 x + b(1 - p)$. Therefore, a solution $p = \bar{p}(x)$ de $g(x, p) = 0$ is the solution of the second-degree polynomial:

$$axp^2 + bp - b = 0$$

let:

$$\bar{p}(x) = \frac{-b + \sqrt{b^2 + 4abx}}{2ax} = \frac{2}{1 + \sqrt{1 + 4\frac{a}{b}x}}$$

which is indeed a decreasing function in x.

A2.4.2. *Exercise 5.2*

1) By performing the same calculations as those which led to equations [5.22], it yields:

$$\begin{cases} \frac{ds}{dt} &= D(S_{in} - s) - \mu(s, x)x \\ \frac{dx}{dt} &= \mu(s, x)x - d(x)x \end{cases}$$

by defining:

$$\mu(s, x) = \bar{p}(x)\mu_u(s) + (1 - \bar{p}(x))\mu_v(s) \quad \text{and} \quad d(x) = \bar{p}(x)D_u + (1 - \bar{p}(x))D_v$$

2) At equilibrium, s^* and x^* verify:

$$s^* = f_1(x^*) := S_{in} - \frac{x^* d(x^*)}{D}$$

and

$$H(s^*, x^*) := \bar{p}(x^*)(\mu_u(s^*) - D_u) + (1 - \bar{p}(x^*))(\mu_v(s^*) - D_v) = 0$$

Given that we have $xd(x) = x\bar{p}(x)(D_u - D_v) + D_v$ and that the function $x \mapsto x\bar{p}(x)$ has been assumed as increasing, it can be deduced that f_1 is decreasing. However, it can immediately be seen that for $s < \min(\lambda_u, \lambda_v)$, we have $H(s, x) < 0$ for all x and that for $s > \max(\lambda_u, \lambda_v)$ we have $H(s, x) > 0$ for all x. We thus necessarily obtain:

$$s^* \in [\min(\lambda_u, \lambda_v), \max(\lambda_u, \lambda_v)].$$

The partial derivatives of H are:

$$\frac{\partial H}{\partial s} = p(x)\mu_u'(s) + (1 - p(x))\mu_v'(s) > 0$$

$$\frac{\partial H}{\partial x} = p'(x)\left[(\mu_u(s) - D_u) - (\mu_v(s) - D_v)\right]$$

When $\lambda_u < \lambda_v$, we have $\mu_u(s) - D_u > 0$ for all $s \in]\lambda_u, \lambda_v]$ and $\mu_v(s) - D < 0$ for all $s \in [\lambda_u, \lambda_v[$. It then follows that $\frac{\partial H}{\partial x} < 0$ for every $s \in [\lambda_u, \lambda_v]$ and $x > 0$. According to the implicit functions theorem, there exists a unique function f_2 such that $H(f_2(x^*), x^*) = 0$ with $f_2'(x) = -\frac{\partial H}{\partial x} / \frac{\partial H}{\partial s} > 0$. Now, s^* verifies $s^* = f_1(x^*) = f_2(x^*)$. Since f_1 and f_2 are, respectively, decreasing and increasing, it can be deduced that there exists at most a solution x^* to the equation $f_1(x^*) = f_2(x^*)$. The corresponding value of s^* is then uniquely defined.

3) When such a positive equilibrium exists, the Jacobian matrix at this equilibrium is written as:

$$J = \begin{bmatrix} -D - \frac{\partial \mu}{\partial s}x & -\frac{\partial \mu}{\partial x}x - d(x) \\ \frac{\partial \mu}{\partial s}x & \frac{\partial \mu}{\partial x}x - d'(x)x \end{bmatrix}$$

from which it is determined that:

$$Tr\, J = -D - \frac{\partial \mu}{\partial s}x + \frac{\partial \mu}{\partial x}x - d'(x)x < 0$$

and:

$$\begin{aligned} Det\, J &= -[D + \tfrac{\partial \mu}{\partial s}x][\tfrac{\partial \mu}{\partial x}x - d'(x)x] + \tfrac{\partial \mu}{\partial s}x[\tfrac{\partial \mu}{\partial x}x + d(x)] \\ &= Dx[d'(x) - \tfrac{\partial \mu}{\partial x}] + x\tfrac{\partial \mu}{\partial s}[xd(x)]' \\ &= Dx\tfrac{\partial \mu}{\partial x}[f_2' - f_1'] > 0 \end{aligned}$$

Therefore, the equilibrium E_1 is locally (exponentially) stable.

4) When $\lambda_u > \lambda_v$, the previous analysis shows that the functions f_1 and f_2 are both decreasing. Therefore, it is possible to have several positive equilibria.

5) The Jacobian matrix for the washout equilibrium is written as:

$$J = \begin{bmatrix} -D & -\mu(S_{in},0) \\ 0 & \mu(S_{in},0) - d(0) \end{bmatrix}$$

whose eigenvalues are $-D$ and $\mu(S_{in},0) - d(0)$. Under the assumptions $\alpha(0,0) = 0$ and $\beta(0) > 0$, note that we necessarily have $\bar{p}(0) = 1$. Then, when $x = 0$ we have $\mu(S_{in},0) = \mu_u(S_{in})$ and $d(0) = D_u$, and the second eigenvalue is thereby negative as soon as $\lambda_u > S_{in}$. The washout equilibrium is (locally) attractive.

A2.5. Appendix exercises

A2.5.1. *Exercise A1.1*

The equation of equilibrium point is:

$$\begin{aligned} 0 &= D(S_{in} - s) - p(s)x \\ 0 &= (q(s) - D_1)x \end{aligned}$$

Consequently, either $x = 0$ and in this case $s = S_{in}$ which gives us the equilibrium E_0, or $q(s) - D_1$, which yields $s = s^*$ defined by this equation, and $x = x^* := \frac{D}{p(s^*)}(S_{in} - s^*)$, that is to say the equilibrium E_1. This equilibrium point only exists if $x^* > 0$, namely, if $S_{in} > s^*$.

A2.5.2. *Exercise A1.2*

In the case $f(x) = ax$, the recursive sequence x_k defined by [A1.4] is given by:

$$x_k = (1 + ah)^k x_0 \tag{A2.17}$$

From [A2.17], it is deduced that the function $x(t,h)$ defined by [A1.5] is:

$$x(t,h) = (1 + ah)^k x_0 \quad \text{for } kh \leq t < (k+1)h \tag{A2.18}$$

Let $t > 0$. From inequalities $kh \leq t < (k+1)h$, we deduce that the number of iterations needed to reach t is the unique integer k defined by:

$$t/h - 1 < k \leq t/h \tag{A2.19}$$

Consequently, from [A2.18] and [A2.19], the upper and lower bounds [A1.6] of $x(t, h)$ can be deduced.

Moreover, we have:

$$\ln(1 + ah) = ah + h\epsilon(h) \quad \text{with} \lim_{h \to 0} \epsilon(h) = 0$$

As a result:

$$\ln\left((1 + ah)^{t/h}\right) = \frac{t}{h} \ln(1 + ah) = \frac{t}{h}(ah + h\epsilon(h)) = t(a + \epsilon(h))$$

Therefrom, it is deduced that:

$$\lim_{h \to 0} \ln\left((1 + ah)^{t/h}\right) = at \implies \lim_{h \to 0}(1 + ah)^{t/h} = e^{at}$$

We also have:

$$\lim_{h \to 0}(1 + ah)^{t/h-1} = \frac{\lim_{h \to 0}(1 + ah)^{t/h}}{\lim_{h \to 0}(1 + ah)} = e^{at}$$

A2.5.3. *Exercise A1.3*

The function $x(t)$ is continuously differentiable and its derivative is equal to:

$$\frac{dx}{dt}(t) = \begin{cases} 3(t - a)^2 & \text{if } t < a \\ 0 & \text{if } a \leq t \leq b \\ 3(t - b)^b & \text{if } t > b \end{cases}$$

Consequently, for all $t \in \mathbb{R}$, it yields that $\frac{dx}{dt}(t) = 3x(t)^{2/3}$. Moreover, $x(0) = 0$ because $a \leq 0 \leq b$.

A2.5.4. *Exercise A1.4*

Because we no longer have a simple linear equation describing the dynamics of $z = s + x$, the procedure is as follows. We denote by δ the smallest of the two quantities D and D_x and to illustrate the idea, we assume that $D_x = \delta + a$ is the largest (we ask the reader to adapt the proof to the case in which D is the largest). Let (s_o, x_o) be an initial condition. Given that for $s > S_{in}$, we always have $\frac{ds}{dt} < 0$, the solution $(s(t), x(t))$ originating from (s_o, x_o) is such that $s(t)$ is in the end smaller than $2S_{in}$. It is thus always possible to assume that s_o is smaller than $2S_{in}$. Since $s \mapsto q(s)$ is zero at 0 and continuous, there is a $s_1 > 0$ such that:

$$s \in [0, s_1] \implies q(s) - D_x \leq \alpha < 0$$

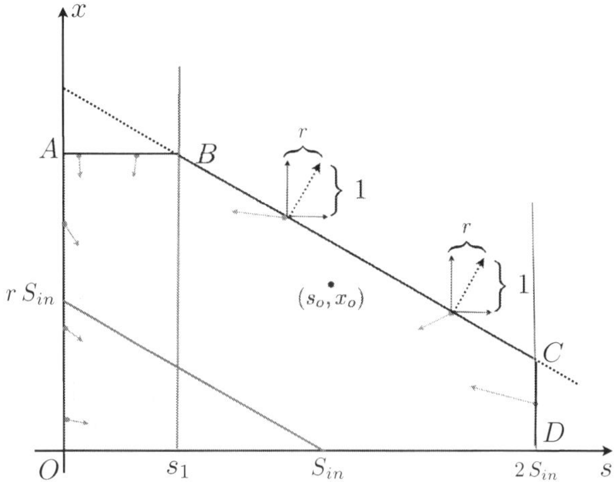

Figure A2.3. *The solutions of [A1.2] are bounded. For a color version of this figure, see www.iste.co.uk/harmand/chemostat.zip*

We choose such an s_1 smaller than $2S_{in}$. In the interval $[s_1, 2S_{in}]$, the quantity $\frac{q(s)}{p(s)}$ is upper bounded, thus r is an upper bound. After drawing a straight line of slope $-r$ that passes above the point (s_o, x_o), we consider the polygon OABCD defined in Figure A2.3. Along the segments OA, AB and CD, the vector field:

$$\begin{bmatrix} \delta(S_{in} - s) - p(s)x \\ (q(s) - (\delta + a))x \end{bmatrix}$$

points toward the interior of the polygon in an obvious manner. The same occurs along BD. In effect, the dot product with the leaving vector normal to BD:

$$P = \left\langle \begin{bmatrix} r \\ 1 \end{bmatrix}, \begin{bmatrix} \delta(S_{in} - s) - p(s)x \\ (q(s) - (\delta + a))x \end{bmatrix} \right\rangle$$

is equal to:

$$P = r\delta(S_{in} - s) - rp(s)x + (q(s) - (\delta + a))x$$
$$P = r\delta S_{in} - r\delta s - rp(s)x + \frac{q(s)}{p(s)}p(s)x - \delta x - ax$$
$$P \leq r\delta S_{in} - r\delta s - rp(s)x + rp(s)x - \delta x - ax \quad \text{according to the definition of } r$$
$$P \leq \delta(rS_{in} - (r\,s + x)) - ax$$
$$P \leq \delta(rS_{in} - (r\,s + x)) \quad\quad\quad \text{because } a \text{ is positive}$$
$$P < 0 \quad\quad\quad\quad \text{BC is above } [(0, r\,S_{in}), (S_{in}, 0)]$$

Therefore, also along BC, the field points inward. The trajectory originating from (s_o, x_o) cannot leave the domain defined by the polygon 0ABCD and as a result it is bounded.

A2.5.5. *Exercise A1.5*

The functions $t \mapsto x(t + T, x_0)$ and $t \mapsto x(t, x_0)$ are both solutions of the system and are equal in $t = 0$. As a result, they are equal for all t. Furthermore, $x(t+T, x_0) = x(t, x_0)$ for all t, that is to say, the solution is T-periodical.

A2.5.6. *Exercise A1.6*

Since $f_i(x^*) \neq 0$, it is assumed that $f_i(x^*) > 0$ (the case $f_i(x^*) < 0$ is addressed in the same manner). Let $a > 0$ such that $f_i(x^*) > a$. Since f is continuous, there exists $\epsilon > 0$ such that:

$$\|x - x^*\| \leq \epsilon \Longrightarrow f_i(x) > a$$

Since $\lim_{t \to +\infty} x(t) = x^*$, there exists $T > 0$ such that:

$$t \geq T \Longrightarrow \|x(t) - x^*\| \leq \epsilon$$

It can be deduced that, for all $t \geq T$, $f_i(x(t)) > a$ and consequently:

$$x_i(t) = x_i(T) + \int_T^t f_i(x(s))ds > x_i(T) + a(t - T)$$

tends to infinity when t tends to infinity, which contradicts: $\lim_{t \to +\infty} x_i(t) = x_i^*$.

A2.5.7. *Exercise A1.7*

Let $x \in \mathcal{D} \subset \mathbb{R} \mapsto f(x) \in \mathbb{R}$ be a vector field and x^* an isolated equilibrium point x^*, that is $f(x^*) = 0$ and there exists $\eta > 0$ such that $f(x) \neq 0$ for $x^* - \eta < x < x^* + \eta$. There are four cases to consider:

1) $f(x) > 0$ for $x^* - \eta < x < x^*$ and $f(x) < 0$ for $x^* < x < x^* + \eta$. In this case, if $x^* - \eta < x_0 < x^*$ (respectively, $x^* < x_0 < x^* + \eta$), the solution $x(t, x_0)$ is strictly increasing and upper bounded by x^* (respectively, strictly decreasing and lower bounded by x^*). Consequently $\lim\limits_{t \to +\infty} x(t, x_0) = x_\infty$ exists and $x^* - \eta < x_\infty < x^* + \eta$. According to proposition A1.4, x_∞ is an equilibrium point. Therefore, $x_\infty = x^*$, because x^* is an isolated equilibrium. Thereof, it can be deduced that x^* is LAS;

2) $f(x) < 0$ for $x^* - \eta < x < x^*$ and $f(x) > 0$ for $x^* < x < x^* + \eta$. In this case, if $x^* - \eta < x_0 < x^*$ (respectively, $x^* < x_0 < x^* + \eta$), the solution $x(t, x_0)$ is strictly decreasing and upper bounded by x^* (respectively, strictly increasing and lower bounded by x^*). Consequently, $\lim\limits_{t \to -\infty} x(t, x_0) = x_\infty$ exists and $x^* - \eta < x_\infty < x^* + \eta$. According to proposition A1.4, x_∞ is an equilibrium point. Therefore, $x_\infty = x^*$, because x^* is an isolated equilibrium. Thereof, it can be deduced that x^* is unstable;

3) $f(x) > 0$ for $x^* - \eta < x < x^*$ and $f(x) > 0$ for $x^* < x < x^* + \eta$ (thus f is canceled without changing sign), using the same method it can be shown that if $x_0 < x^*$, then $\lim\limits_{t \to +\infty} x(t, x_0) = x^*$ and if $x_0 > x^*$, then $\lim\limits_{t \to -\infty} x(t, x_0) = x^*$. As a result, x^* is attractive on the left side and repulsive on the right side. It is thus unstable;

4) $f(x) < 0$ for $x^* - \eta < x < x^*$ and $f(x) < 0$ for $x^* < x < x^* + \eta$ (thus f is canceled out without changing sign), using the same method it can be shown that if $x_0 < x^*$, then $\lim\limits_{t \to -\infty} x(t, x_0) = x^*$ and if $x_0 > x^*$, then $\lim\limits_{t \to +\infty} x(t, x_0) = x^*$. Consequently, x^* is repulsive on the left side and attractive on the right side. It is thus unstable.

A2.5.8. *Exercise A1.8*

From $\frac{dy_2}{dt} = \lambda y_2$, it can be deduced that: $y_2(t) = c_2 e^{\lambda t}$. Define $y_1(t) = z(t)e^{\lambda t}$. It follows that $\frac{dy_1}{dt} = \frac{dz}{dt} e^{\lambda t} + \lambda z(t)e^{\lambda t}$. Since $\frac{dy_1}{dt} = \lambda y_1 + y_2$, it can be deduced that:

$$\frac{dz}{dt} e^{\lambda t} + \lambda z(t)e^{\lambda t} = \lambda z(t)e^{\lambda t} + c_2 e^{\lambda t}$$

Furthermore, $\frac{dz}{dt} = c_2$ and therefore $z(t) = c_2 t + c_1$.

A2.5.9. *Exercise A1.9*

If x^* is LES, then there exist $\alpha > 0$, $\beta > 0$ and $\eta > 0$ such that for any initial condition x_0, we have:

$$\|x_0 - x^*\| < \eta \implies \text{for all } t \geq 0, \|x(t, x_0) - x^*\| \leq \beta\|x_0 - x^*\|e^{-\alpha t}$$

Consequently for:

$$\|x_0 - x^*\| < \eta \implies \lim_{t \to +\infty} x(t, x_0) = x^*$$

which implies that x^* is attractive. Moreover, given that $\varepsilon > 0$, if $\|x_0 - x^*\| < \delta$ with $\delta = \varepsilon/\beta$, for all $t \geq 0$, we get:

$$\|x(t, x_0) - x^*\| \leq \beta\|x_0 - x^*\|e^{-\alpha t} < \varepsilon$$

Therefrom, it can be deduced that x^* is stable. As it has also been shown that it is attractive, it is therefore LAS.

In order to show that LAS does not imply LES, consider the scalar equation:

$$\frac{dx}{dt} = -x^3$$

According to exercise A1.7, the origin is LAS (and even GAS) because the function $x \mapsto -x^3$ cancels out and changes sign, from positive to negative when x crosses the origin and is increasing. To show that the origin is not LES, we integrate the equation. The solution of initial condition $x_0 \neq 0$ is written as:

$$x(t, x_0) = \frac{x_0}{\sqrt{1 + 2x_0^2 t}}, \qquad \text{for } t \in \left(-1/2x_0^2, +\infty\right)$$

If the origin is LES, there exist $\alpha > 0$, $\beta > 0$ and $\eta > 0$ such that for any initial condition x_0 verifying $\|x_0 - x^*\| < \eta$ and every $t \geq 0$, we have:

$$|x(t, x_0)| = \frac{|x_0|}{\sqrt{1 + 2x_0^2 t}} \leq \beta|x_0|e^{-\alpha t}$$

It can be deduced that for all $t \geq 0$, we have:

$$\sqrt{1 + 2x_0^2 t} \geq \frac{e^{\alpha t}}{\beta}$$

which is impossible because an exponential function increases more quickly to infinity than the root function.

A2.5.10. *Exercise A1.10*

The Jacobian matrix of [A1.2] in (s, x) is:

$$J = \begin{pmatrix} -D - p'(s)x & -p(s) \\ q'(s)x & q(s) - D_1 \end{pmatrix}$$

In E_0, this matrix is equal to:

$$J_0 = \begin{pmatrix} -D & -p(S_{in}) \\ 0 & q(S_{in}) - D_1 \end{pmatrix}$$

The eigenvalues are $-D$ and $q(S_{in}) - D_1 > 0$. Consequently, according to proposition A1.6, E_0 is a saddle. For the existence of the invariant varieties for E_0, theorem A1.6 is used. It is known that the stable variety is unique and it is tangent to the axis $x = 0$ which is the eigensubspace E^s at the equilibrium point E_0. Since we know that the axis $x = 0$ is invariant and that solutions in this axis tend to E_0, it is indeed the stable variety.

A2.5.11. *Exercise A1.11*

At E_1, the Jacobian matrix of [A1.2] is equal to:

$$J = \begin{pmatrix} -D - p'(s^*)x^* & -p(s^*) \\ q'(s^*)x^* & 0 \end{pmatrix}$$

It follows that:

$$\det(J) = p(s^*)q'(s^*)x^*, \qquad \mathrm{Tr}(J) = -D - p'(s^*)x^* = \pi'(s^*)p(s^*)$$

Consequently, $\det(J) > 0$ and $\mathrm{Tr}(J) < 0$ if and only if $q'(s^*) > 0$ and $\pi'(s^*) < 0$.

A2.5.12. *Exercise A1.12*

Since the equation $q(s) = D_1$ admits two solutions s_1^* and s_2^*, and two only, such that $0 < s_1^* < s_2^* < S_{in}$, and that $q(s) > 0$ for every $s > 0$, we deduce that $q(S_{in}) < D_1$. As a result, see the solution of exercise A1.10, the washout E_0 is LES. Following proposition A1.6 and using exercise A1.11, E_2 is a saddle because $q'(s_2^*)x_2^* < 0$ and E_1 is a sink because $q'(s_1^*) > 0$ and $\pi'(s_1^*) < 0$.

A2.5.13. *Exercise A1.13*

If x_0 is an equilibrium point, then $x(t, x_0) = x_0$ for every $t \in \mathbb{R}$. Thereby, $\lim_{k \to \infty} x(t_k, x_0) = x_0$ for any sequence t_k. It can be deduced that the limit set is $\{x_0\}$. For a periodical solution $x(t)$ of period T, we have $x(t) = x(t + kT)$ for any integer k. As a result, $x(t_k)$, where $t_k = t + kT$ tends to $x(t)$ when k tends to infinity. We have thus shown that any point $x(t)$ of the periodic orbit is a point of the ω-limit set. If $b \in \gamma(a)$, then there exists τ such that $b = x(\tau, a)$, so that:

$$x(t, b) = x(t, x(\tau, a)) = x(t + \tau, a)$$

Therefore, the functions $x(t, b)$ and $x(t, a)$ have the same possible limits.

A2.5.14. *Exercise A1.14*

In polar coordinates, [A1.23] is written as:

$$\begin{aligned}
\frac{dr}{dt} &= r(1 - r) \\
\frac{d\theta}{dt} &= r\beta |\sin\theta|
\end{aligned} \qquad \text{[A2.20]}$$

The equilibrium points are given by $r = 0$, corresponding to $E_0 = (0, 0)$, as well as $\sin\theta = 0$ and $r = 1$, corresponding to $E_1 = (1, 0)$ and $E_2 = (1, 0)$. The axis $\sin\theta = 0$, corresponding to $x_2 = 0$ is invariant because $\sin\theta = 0$ implies $\frac{d\theta}{dt} = 0$. Similarly, the circle $r = 1$ is invariant because $r = 1$ implies $\frac{dr}{dt} = 0$. Given that there are no other equilibrium points, the semi-circles $\gamma_1 = \{r = 1, x_2 > 0\}$ and $\gamma_2 = \{r = 1, x_2 < 0\}$ are orbits. Consider a solution with initial condition $x_2(0) > 0$. If $r(0) < 1$, then the solution remains within the upper half-disk and since $r(t)$ is increasing and upper bounded by 1, and $\theta(t)$ is increasing and upper bounded by π, the limits:

$$\lim_{t \to +\infty} r(t) = r_\infty > 0, \qquad \lim_{t \to +\infty} \theta(t) = \theta_\infty > 0$$

exist and are strictly positive. According to proposition A1.4, $(r_\infty, \theta_\infty)$ is an equilibrium point of [A2.20] belonging to the upper half-disk. Since $E_2 = (-1, 0)$ is the only equilibrium point verifying $\theta > 0$ and $r > 0$, it is deduced that $(r_\infty, \theta_\infty) = E_2$. Similarly, if $r(0) > 1$, then $r(t)$ is decreasing and lower bounded by 1 and it is therefore concluded that it admits a limit. The same argument shows that if $x_2(0) < 0$, all solutions converge to E_1.

A2.5.15. *Exercise A1.15*

Since $s \mapsto q(s)$ is strictly increasing and $q(S_{in}) > D_1$, $E_1 = (s^*, x^*)$ is the only positive equilibrium. According to proposition A1.6 and using the solution of exercise A1.11, E_1 is a source because:

$$\det(J) = p(s^*)q'(s^*)x^* > 0, \qquad \mathrm{Tr}(J) = \pi'(s^*)p(s^*) > 0$$

Let (s_0, x_0) be an initial condition such that $x_0 > 0$. According to Exercise A1.4, the solutions are positively bounded. Consequently, the limit set ω of the orbit $\gamma(s_0, x_0)$ is non-empty and included in $\mathcal{D}$. Following proposition A1.8, this limit set cannot contain the equilibrium point E_0. It also cannot contain the point E_1 because it is a source. According to the Poincaré–Bendixon theorem, the ω-limit set is a periodic orbit. This periodic orbit must contain an equilibrium point in its interior. Given that it may not contain the washout E_0 in its interior, it necessarily surrounds E_1, since it is the only positive equilibrium.

A2.5.16. *Exercise A1.16*

The derivative of V along the trajectories of [A1.28] is:

$$\dot{V}(s, x) = \frac{\partial V}{\partial s}\frac{ds}{dt} + \frac{\partial V}{\partial x}\frac{dx}{dt} = x^* \frac{q(s) - D_1}{D(S_{in} - s)}\frac{ds}{dt} + \frac{x - x^*}{x}\frac{dx}{dt}$$

Using [A1.28] yields:

$$\dot{V}(s, x) = x^* \frac{q(s) - D_1}{D(S_{in} - s)}\left[D(S_{in} - s) - p(s)x\right] + (x - x^*)(q(s) - D_1)$$

After simplification of terms $x^*(q(s) - D_1)$ and using [A1.29] and $x^* = P(s^*)$, we obtain:

$$\dot{V}(s, x) = x(q(s) - D_1)\left[1 - \frac{x^*}{\pi(s)}\right] = x(q(s) - D_1)\frac{\pi(s) - \pi(s^*)}{\pi(s^*)}$$

According to [A1.31], $V'(s, x)$ is strictly negative for $x > 0$ and $0 < s < S_{in}$ such that $s \neq s^*$. It follows that:

$$\mathcal{V} = \{\dot{V}(s, x)\} = \{s = s^* \text{ or } x = 0\}$$

The largest invariant set contained in $\mathcal{V}$ is constituted of the two equilibrium points $M = \{E_0, E_1\}$. Using theorem A1.13, we can deduce that the ω-limit set is included in A. The solutions tend to M. Since it is connected and non-empty, it can only be equal to $\{E_0\}$ or to $\{E_1\}$. However, given that $x(0) > 0$, the solutions can only tend to E_1.

Bibliography

[ABD 16] ABDELLATIF N., FEKIH-SALEM R., SARI T., "Competition for a single resource and coexistence of several specises in the chemostat", *Mathematical Biosciences & Engineering*, vol. 13, pp. 631–652, 2016.

[ADA 15] ADAMBERG K., VALGEPEA K., VILU R., "Advanced continuous cultivation methods for systems microbiology", *Microbiology-Sgm*, vol. 161, pp. 1707–1719, 2015.

[AND 68] ANDREWS J., "A mathematical model for the continuous culture of microorganisms utilizing inhibitory substrates", *Biotechnology & Bioengineering*, vol. 10, no. 6, pp. 707–723, 1968.

[ANG 04] ANGELI D., DE LEENHEER P., SONTAG E., "A small gain theorem for almost global convergence of monotone systems", *Systems and Control Letters*, vol. 52, pp. 407–414, 2004.

[ANO 88] ANOSOV D., ARNOLD V., *Dynamical Systems I, Ordinary Differential Equations and Smooth Dynamical Systems*, Springer–Verlag, 1988.

[ARD 89] ARDITI R., GINZBURG L., "Coupling in Predator-Prey Dynamics: Ratio-Dependence", *Journal of Theoretical Biology*, vol. 139, pp. 311–328, 1989.

[ARD 92] ARDITI R., SAIAH H., "Empirical evidence of the role of heterogeneity in ratio-dependent consumption", *Ecology*, vol. 73, no. 5, pp. 1544–1551, 1992.

[ARD 12] ARDITI R., GINZBURG L., *How Species Interact: Altering the Standard View on Thophic Ecology*, Oxford University Press, Oxford, 2012.

[ARI 03] ARINO J., PILYUGIN S., WOLKOWICZ G., "Considerations on yield, nutrient uptake, cellular growth, and competition in chemostat models", *Canadian Applied Mathematics Quarterly*, vol. 11, pp. 107–142, 2003.

[ARI 77] ARIS R., HUMPHREY A., "Dynamics of a Chemostat in which Two Organisms Compete for a Common Substrate", *Biotechnology & Bioengineering*, vol. 19, no. 9, pp. 1375–1386, 1977.

[ARM 80] ARMSTRONG R., MCGEHEE R., "Competitive exclusion", *American Naturalist*, vol. 115, pp. 151–170, 1980.

[ARN 74] ARNOLD V., *Equations différentielles ordinaires*, Editions Mir, Moscou, 1974.

[AUT 10] AUTO, "Software for continuation and bifurcation problems in ordinary differential equations", http://indy.cs.concordia.ca/auto/, 2010.

[BAL 99] BALLYK M., SMITH H., "A model of microbial growth in a plug flow reactor with wall attachment", *Mathematical Biosciences*, vol. 158, pp. 95–126, 1999.

[BAL 08] BALLYK M., JONES D., SMITH H., "The biofilm model of Freter: a review", in *Structured Population Models in Biology and Epidemiology*, Springer, Berlin, 2008.

[BAS 90] BASTIN G., DOCHAIN D., *On-Line Estimation and Control of Bioreactors*, Elsevier Science Publishers, Amsterdam, 1990.

[BER 95] BERLIN A., KISLENKO V., "Kinetic models of suspension flocculation by polymers", *Colloids and Surfaces A: Physicochemical and Engineering Aspects*, vol. 104, pp. 67–72, 1995.

[BER 01] BERNARD O., HADJ-SADOK Z., DOCHAIN D. *et al.*, "Dynamical model development and parameter identification for an anaerobic wastewater treatment process", *Biotechnology & bioengineering*, vol. 75, no. 4, pp. 424–438, 2001.

[BUT 85] BUTLER G., WOLKOWICZ G., "A Mathematical Model of the Chemostat with a General Class of Functions Describing Nutrient Uptake", *SIAM Journal on Applied Mathematics*, vol. 45, pp. 137–151, 1985.

[BUT 86] BUTLER G., WOLKOWICZ G., "Predator-mediated competition in the chemostat", *Journal of Mathematical Biology*, vol. 24, no. 2, pp. 167–191, 1986.

[CHO 59] CHOLETTE A., CLOUTIER L., "Mixing efficiency determinations for continuous flow systems", *The Canadian Journal of Chemical Engineering*, vol. 37, pp. 105–110, 1959.

[CHO 60] CHOLETTE A., BLANCHET J., "Performance of flow reactors at various levels of mixing", *The Canadian Journal of Chemical Engineering*, vol. 38, pp. 1–18, 1960.

[CHO 61] CHOLETTE A., BLANCHET J., "Optimum performance of combined flow reactors under adiabatic conditions", *The Canadian Journal of Chemical Engineering*, vol. 39, pp. 192–198, 1961.

[COO 73] COOPER D., COPELAND B., "Responses of a continuous series of estuarine microecosystems to point-source input variations", *Ecological Monographs*, vol. 43, pp. 213–236, 1973.

[COP 96] COPIN-MONTEGUT G., *Le système des carbonates*, Institut Océanographique de Paris, 1996.

[COS 79] COSTE J., PEYRAUD J., COULLET P., "Asymptotic behaviors in the dynamics of competing species", *SIAM Journal on Applied Mathematics*, vol. 36, no. 3, pp. 516–543, SIAM, 1979.

[COS 95] COSTERON J., "Overview of microbial biofilms", *Journal of Industrial Microbiology*, vol. 15, pp. 137–140, 1995.

[DEL 03] DE LEENHEER P., LI B., SMITH H., "Competition in the chemostat: Some remarks", *Canadian Applied Mathematics Quarterly*, vol. 11, no. 3, pp. 229–248, 2003.

[DEL 06] DE LEENHEER P., ANGELI D., SONTAG E.D., "Crowding effects promote coexistence in the chemostat", *Journal of Mathematical Analysis and Applications*, vol. 319, no. 1, pp. 48–60, 2006.

[DEM 06] DEMAILLY J.P., *Analyse numérique et équations différentielles*, EDP Sciences, Les Ulis, 2006.

[DER 75] DER YANG R., HUMPHREY A., "Dynamic and steady state studies of phenol biodegradation in pure and mixed cultures", *Biotechnology & Bioengineering*, vol. 17, no. 8, pp. 1211–1235, 1975.

[ESC 05] ESCUDIÉ R., CONTE T., STEYER J.P. *et al.*, "Hydrodynamic and biokinetic models of an anaerobic fixed-bed reactor", *Process Biochemistry*, vol. 40, no. 7, pp. 2311–2323, 2005.

[FEK 13a] FEKIH-SALEM R., Modèles mathématiques pour la compétition et la coexistence des espèces microbiennes dans un chémostat, Thesis, University of Montpellier 2 and Tunis El Manar University, 2013.

[FEK 13b] FEKIH-SALEM R., HARMAND J., LOBRY C. *et al.*, "Extensions of the chemostat model with flocculation", *Journal of Mathematical Analysis and Applications*, vol. 397, pp. 292–306, 2013.

[FEK 16] FEKIH-SALEM R., RAPAPORT A., SARI T., "Emergence of coexistence and limit cycles in the chemostat model with flocculation for a general class of functional responses", *Applied Mathematical Modelling*, vol. 40, pp. 765–7677, 2016.

[FEK 17] FEKIH-SALEM R., LOBRY C., SARI T., "A density-dependent model of competition for one resource in the chemostat", *Mathematical Biosciences*, vol. 286, pp. 104 122, 2017.

[FRE 83] FRETER R., BRICKNER H., FEKETE J. *et al.*, "Survival and implantation of Escherichia coli in the intestinal tract", *Infection and Immunity*, vol. 39, pp. 686–703, 1983.

[GAU 32] GAUSE G.F., "Experimental studies on the struggle for existence", *Journal of Experimental Biology*, vol. 9, pp. 389-402, 1932.

[GAU 03] GAUSE G.F., *The Struggle for Existence*, Dover Phoneix Editions, 2003.

[GOD 66] GODEMENT R., *Cours d'algèbre*, Hermann, Paris, 1966.

[GRA 72] GRADY JR C., HARLOW L.J., RIESING R.R., "Effects of growth rate and influent substrate concentration on effluent quality from chemostats containing bacteria in pure and mixed culture", *Biotechnology & Bioengineering*, vol. 14, pp. 391–410, 1972.

[GRA 75] GRADY JR C., WILLIAMS D.R., "Effects of influent substrate concentration on the kinetics of natural microbial populations in continuous culture", *Water Research*, vol. 9, pp. 171–180, 1975.

[GRO 07] GROGNARD F., MAZENC F., RAPAPORT A., "Polytopic Lyapunov functions for persistence analysis of competing species", *Discrete and Continuous Dynamical Systems – Series B*, vol. 8, no. 1, pp. 73-93, 2007.

[GUC 13] GUCKENHEIMER J., HOLMES P.J., *Nonlinear Oscillations, Dynamical Systems, and Bifurcations of Vector Fields*, Springer Science & Business Media, 2013.

[HAE 07] HAEGEMAN B., LOBRY C., HARMAND J., "Modeling bacteria flocculation as density-dependent growth", *AIChE Journal*, vol. 53, no. 2, pp. 535–539, 2007.

[HAE 08] HAEGEMAN B., RAPAPORT A., "How flocculation can explain coexistence in the chemostat", *Journal Biological Dynamics*, vol. 2, pp. 1–13, 2008.

[HAH 67] HAHN W., *Stability of Motion*, Springer-Verlag, 1967.

[HAL 30] HALDANE J.S.H., *Enzymes*, Longmans Green and Co, London, 1930.

[HAL 69] HALE J., *Ordinary Differential Equations*, Wiley-Interscience, 1969.

[HAN 80] HANSEN S., HUBBELL S., "Single-nutrient microbial competition: qualitative agreement between experimental and theoretically forecast outcomes", *Science*, vol. 28, pp. 1491–1493, 1980.

[HAR 07] HARMAND J., GODON J., "Density-dependent kinetics models for a simple description of complex phenomena in macroscopic mass-balance modeling of bioreactors", *Ecological Modelling*, vol. 200, nos. 3-4, pp. 393–402, 2007.

[HAR 08] HARMAND J., RAPAPORT A., DOCHAIN D., LOBRY C., "Microbial ecology and bioprocess control: Opportunities and challenges", *Journal of Process Control*, vol. 18, no. 9, pp. 865–875, 2008.

[HEF 09] HEFFERNAN B., MURPHY C., CASEY E., "Comparison of planktonic and biofilm cultures of Pseudomonas fluorescens DSM 8341 cells grown on fluoroacetate", *Applied Environmental Microbiology*, vol. 75, pp. 2899–2907, 2009.

[HIR 74] HIRSCH M.W., SMALE S., *Differential Equations, Dynamical Systems and Linear Algebra*, Academic Press, New York, 1974.

[HIR 77] HIRSCH M.W., SHUB M., PUGH C.C., *Invariant Manifolds*, Springer-Verlag, 1977.

[HIR 82] HIRSCH M.W., "Systems of differential equations that are competitive or cooperative: I. limit sets", *SIAM Journal on Mathematical Analysis*, vol. 13, no. 2, pp. 167–179, 1982.

[HIR 85] HIRSCH M.W., "Systems of differential equations that are competitive or cooperative II: Convergence almost everywhere", *SIAM Journal on Mathematical Analysis*, vol. 16, no. 3, pp. 423–439, 1985.

[HIR 88] HIRSCH M.W., "Systems of differential equations that are competitive or cooperative: III. Competing species", *Nonlinearity*, vol. 1, no. 1, p. 51, 1988.

[HOS 05] HOSKISSON P.A., HOBBS G., "Continuous culture - making a comeback?", *Microbiology-Sgm*, vol. 151, pp. 3153-3159, 2005.

[HSU 77] HSU S., HUBBELL S., WALTMAN P., "A mathematical theory for single-nutrient competition in continuous cultures of microorganisms", *SIAM Journal on Applied Mathematics*, vol. 32, pp. 366–383, 1977.

[HSU 78] HSU S., "Limiting behavior for competing species", *SIAM Journal on Applied Mathematics*, vol. 34, pp. 760–763, 1978.

[HSU 05] HSU S., "A survey of constructing Lyapunov functions for mathematical models in population biology", *Taiwanese Journal of Mathematics*, vol. 9, no. 2, pp. 151–173, 2005.

[IWA 06] IWA TASK GROUP ON BIOFILM MODELING, *Mathematical Modeling of Biofilms*, IWA Publishing, London, 2006.

[JES 04] JESSUP C.M., KASSEN R., FORDE S.E. *et al.*, "Big questions, small world: microbial model systems in ecology", *Trends in Ecology and Evolution*, vol. 19, no. 4, pp. 189–197, 2004.

[JOH 11] JOHNSON K.A., GOODY R.S., "The original Michaelis constant: translation of the 1913 Michaelis-Menten paper", *Biochemistry*, vol. 50, no. 39, pp. 8264–8269, 2011.

[JON 03] JONES D., KOJOUHAROV H., LE D. *et al.*, "The Freter model: A simple model of biofilm formation", *Journal of Mathematical Biology*, vol. 47, pp. 137–152, 2003.

[JOS 00] JOST C., "Predator-prey theory: hidden twins in ecology and microbiology", *OIKOS*, vol. 90, no. 1, pp. 202–208, 2000.

[KUA 89] KUANG Y., "Limit cycles in a chemostat-related model", *SIAM, Journal on Applied Mathematics*, vol. 49, no. 6, pp. 1759–1767, 1989.

[LEF 63] LEFSCHETZ S., *Differential Equations: Geometric Theory*, Wiley-Interscience, New York, 1963.

[LI 98] LI B., "Global asymptotic behavior of the chemostat: General response functions and differential removal rates", *SIAM Journal on Applied Mathematics*, vol. 59, pp. 411–422, 1998.

[LOB 05] LOBRY C., MAZENC F., RAPAPORT A., "Persistence in ecological models of competition for a single resource", *Comptes Rendus Mathématique*, vol. 340, no. 3, pp. 199–204, 2005.

[LOB 06a] LOBRY C., HARMAND J., "A new hypothesis to explain the coexistence of n species in the presence of a single resource", *Comptes Rendus Biologies*, vol. 329, no. 1, pp. 40–46, 2006.

[LOB 06b] LOBRY C., RAPAPORT A., MAZENC F., "Sur un modèle densité-dépendant de compétition pour une ressource", *Comptes Rendus Biologies*, vol. 329, no. 2, pp. 63–70, 2006.

[LOB 09] LOBRY C., SARI T., RAPAPORT A., "Stability loss delay in the chemostat with a slowly varying washout rate", *Proceedings MATHMOD*, Vienna, Austria, pp. 1582–1586, 2009.

[LOB 13] LOBRY C., "La compétition dans le chémostat", in *Travaux En Cours 81: Des Nombres et des Mondes*, Herman, Paris, 2013.

[LOV 79] LOVITT R., WIMPENNY J., "The gradostat: a tool for investigating microbial growth and interactions insolute gradients", *Society for General Microbiology Quarterly*, vol. 6, no. 80, 1979.

[MAR 56] MARKUS L., "Asymptotic autonomous differential systems", *Annals of Mathematic Studies*, vol. 36, pp. 17–29, 1956.

[MAR 12] MARSDEN J.E., MCCRACKEN M., *The Hopf Bifurcation and its Applications*, Springer Science & Business Media, Heildelberg, 2012.

[MAY 75] MAY R.M., LEONARD W.J., "Nonlinear aspects of competition between three species", *SIAM Journal on Applied Mathematics*, vol. 29, no. 2, pp. 243–253, 1975.

[MIS 95] MISCHAIKOW K., SMITH H., THIEME H.R., "Asymptotically autonomous semi flows: chain recurrence and Lyapunov functions", *Transactions of American Mathematical Society*, vol. 347, no. 5, pp. 1669–1685, 1995.

[MON 50] MONOD J., "La technique de culture continue : Théorie et applications", *Annales de l'Institut Pasteur*, vol. 79, pp. 390–410, 1950.

[NEM 60] NEMYTSKII V.V., STEPANOV V., *Qualitative Theory of Differential Equations*, Princeton University Press, New York, 1960.

[NOV 50] NOVICK A., SZILARD L., "Experiments with the chemostat on spontaneous mutations of bacteria", *Proceedings of the National Academy of Sciences*, vol. 36, no. 12, pp. 708–719, 1950.

[PER 13] PERKO L., *Differential Equations and Dynamical Systems*, vol. 7, Springer Science & Business Media, Heildelberg, 2013.

[PIL 99] PILYUGIN S., WALTMAN P., "The simple chemostat with wall growth", *SIAM Journal on Applied Mathematics*, vol. 59, pp. 1552–1572, 1999.

[PIL 03] PILYUGIN S., WALTMAN P., "Multiple limit cycles in the chemostat with variable yields", *Mathematical Biosciences*, vol. 182, pp. 151–166, 2003.

[RAP 08] RAPAPORT A., HARMAND J., "Biological control of the chemostat with nonmonotonic response and different removal rates", *Mathematical Biosciences & Engineering*, vol. 5, no. 3, pp. 539–547, 2008.

[SAR 11] SARI T., MAZENC F., "Global dynamics of the chemostat with different removal rates and variable yields", *Mathematical Biosciences & Engineering*, vol. 8, no. 3, pp. 827–840, 2011.

[SAR 13] SARI T., "Competitive Exclusion for Chemostat Equations with Variable Yields", *Acta Applicandae Mathematicae*, vol. 123, no. 1, pp. 201–219, 2013.

[SCH 16] SCHOLARPEDIA, "Andronov-Hopf bifurcation", http://www.scholarpedia.org/article/Andronov-Hopf_bifurcation, 2016.

[SMA 76] SMALE S., "On the differential equations of species in competition", *Journal of Mathematical Biology*, vol. 3, no. 1, pp. 5–7, 1976.

[SMI 95] SMITH H.L., WALTMAN P., *The Theory of the Chemostat: Dynamics of Microbial Competition*, vol. 13, Cambridge University Press, Cambridge, 1995.

[SPI 55] SPICER C., "The theory of bacterial constant growth apparatus", *Biometrics*, vol. 11, pp. 225–230, June 1955.

[STE 00] STEMMONS E., SMITH H., "Competition in a chemostat with wall attachment", *SIAM Journal on Applied Mathematics*, vol. 61, pp. 567–595, 2000.

[TAN 97] TANG B., SITOMER A., JACKSON T., "Population dynamics and competition in chemostat models with adaptive nutrient uptake", *Journal of Mathematical Biology*, vol. 35, pp. 453–479, 1997.

[THI 92] THIEME H.R., "Convergence results and a Poincaré-Bendixon trichotomy for asymptotically autonomous differential equations", *Journal of Mathematical Biology*, vol. 30, pp. 755–763, 1992.

[THI 94] THIEME H.R., "Asymptotic autonomous differential equations in the plane", *Rocky Mountain Journal of Mathematics*, vol. 24, pp. 351–380, 1994.

[THO 99] THOMAS D., JUDD S., FAWCETT N., "Flocculation modelling: a review", *Water Research*, vol. 33, pp. 1579–1592, 1999.

[TRU 09] TRUONG-MEYER X., "Modélisation en génie des procédés", Série Procédés chimie - bio - agro - Opérations unitaires - Génie de la réaction chimique, *Techniques de l'Ingénieur*, vol. J1021, 2009.

[TSU 72] TSUCHIYA H., DRAKE J., JOST J. *et al.*, "Predator-prey interactions of Dictyostelium discoideum and Escherichia coli in continuous culture", *Journal of Bacteriology*, vol. 110, no. 3, pp. 1147–1153, 1972.

[WAD 16] WADE M., HARMAND J., BENYAHIA B. *et al.*, "Perspectives in Mathematical Modelling for Microbial Ecology", *Ecological Modelling*, vol. 321, pp. 64–74, 2016.

[WOL 89] WOLKOWICZ G., "Successful invasion of a food web in a chemostat", *Mathematical Biosciences*, Elsevier, vol. 93, no. 2, pp. 249–268, 1989.

[WOL 92] WOLKOWICZ G., LU Z., "Global dynamics of a mathematical model of competition in the chemostat: General response functions and differential death rates", *SIAM Journal on Applied Mathematics*, vol. 52, pp. 222–233, 1992.

[ZIV 13] ZIV N., BRANDT N.J., GRESHAM D., "The Use of Chemostats in Microbial Systems Biology", *Jove-Journal of Visualized Experiments*, vol. 80, no. 10, 2013.

2016

SOUSTELLE Michel

Chemical Thermodynamics Set

Volume 5 – Phase Transformations

Volume 6 – Ionic and Electrochemical Equilibria

Volume 7 – Thermodynamics of Surfaces and Capillary Systems

2015

SOUSTELLE Michel

Chemical Thermodynamics Set

Volume 1 – Phase Modeling Tools

Volume 2 – Modeling of Liquid Phases

Volume 3 – Thermodynamic Modeling of Solid Phases

Volume 4 – Chemical Equilibria

2014

DAL PONT Jean-Pierre, AZZARO-PANTEL Catherine
New Approaches to the Process Industries: The Manufacturing Plant of the Future

HAMAIDE Thierry, DETERRE Rémi, FELLER Jean-François
Environmental Impact of Polymers

2012

DAL PONT Jean-Pierre
Process Engineering and Industrial Management

2011

SOUSTELLE Michel
An Introduction to Chemical Kinetics

2010

SOUSTELLE Michel
Handbook of Heterogenous Kinetics

Printed and bound by CPI Group (UK) Ltd, Croydon, CR0 4YY